D0906070

Farm Fever

FARM FEVER

How to Buy Country Land and Farm It—Part Time or Full Time

Jerry Baker & Dan Kibbie

Funk & Wagnalls New York

Designed by Joy Chu

Manufactured in the United States of America

Library of Congress Cataloging in Publication Data

Baker, Jerry.
Farm fever.

Includes index.
1. Agriculture—Handbook, manuals, etc.
2. Agriculture—United States.
I. Kibbie, Dan, joint author. II. Title.
S501.2.B34 1977 630 76-30861
ISBN 0-308-10299-1
1 3 5 7 9 10 8 6 4 2

Contents

Foreword

Since you have already gone to the trouble to pick this book, it is quite possible that you've already been bitten by the bug that's responsible for one of America's most popular "diseases"—*farm fever!*

Farm fever is an affliction that has struck down many an urbanite right in the middle of his chasing the almighty buck, or his own small share of that carrot called "corporate security." It's a disease that is indiscriminate in whom it attacks, selecting some of society's least likely victims.

I've become something of a local expert in spotting the telltale symptoms. A sure sign is when one of my neighbors begins to wear an old straw hat on the weekends. Pretty soon that hat is just a small part of his "rustic ensemble" which includes faded coveralls and a red bandana whenever he's out riding his power lawn mower.

Then, one day, when his gardening and lawn chores are finished, he's sure to come over to my place and ask me to join him for a few pulls on a jug of hard cider or "corn likker." It's sometimes a little hard to understand the invitation, though, because of the blade of timothy between his teeth and the chaw of tobacco tucked up in his cheek.

After a couple of swigs from the old brown jug, the farm fever will bring a flush to his fluorescent-pale face and, in between country colloquialisms and sage sayings, he will turn the conversation to that all-important subject which proves that he has been bitten.

"Yep, Jerry, I've been giving a lot of thought lately to the idea of moving the family farther out into the country and buying a little farm. We could work the land and raise our own food. It'll be a great thing for all of us, especially the kids. I can't wait to start breathing that fresh air and eating that fresh, wholesome grub!"

Now my neighbor, whose name is Clement Clements III, works in downtown Detroit as a securities salesman, so I can't help but question his sudden interest in agricultural pursuits since it doesn't seem to jibe with his background, experience, or what I know of his family.

"But, Clement," I start to say . . .

"Call me Clem, Jerry."

"Well, Clem," I begin, feeling slightly awkward, "farming is a great calling and all that, but are you sure it's the best thing for you and your family at this time of your life? Won't your oldest boy be going to college in a year or two?"

I don't really know Clement's kids, but I've seen the oldest boy and girl tooling around town in their shiny little sports cars. And my kids, whom I sometimes think are slightly spoiled, use them as examples when they give their reasons for getting out of chores or staying out late.

"Well, Jerry, I haven't really called a family powwow about this yet, but I really feel that buying a farm and getting back to basics is one of the best things that could happen to all of us right now. Just quitting my job and getting out of the System will be a big boost for my family. The way we've been living is too phony and unproductive."

My wife, Ilene, who sometimes works with Clement's wife on community projects, says that their marriage has been on shaky ground for some time now. It seems pretty obvious that one of the reasons he wants to move to the country is to straighten out his family affairs. But any experienced member of a farm family could tell him that life on a farm isn't a cure-all. It's what you make it. I wonder whether Clem has thought about this?

"Clem, what you are thinking of doing is a mighty big step, one that should have the wholehearted support of every member of your family. Buying a farm is no sure way to bring your family together or make your marriage work. Farm life can be tough and lonely. It should be something that everyone wants to do, or chances are that it won't work out. The country is no place to 'get away to.' I'd suggest that you and your family talk this over thoroughly. Then, take some time, perhaps several months, to explore the idea of acquiring a farm and working it before you go ahead and change your life-style so dramatically. Each member of your family should weigh the good aspects of living on a farm against the bad ones. Some of your best sources will be: the United States Department of Agriculture (USDA), reliable farm realtors, farmers, newspapers and farm journals, professional farmer organizations, county agents, local residents of any given farm community, and people who have given up farming and moved to the city. You should make this a family project and have a complete agreement and a thorough idea of what you are going to do before you buy any farm property or dive into any type of farm operation."

This kind of straight talk sometimes has the effect of sobering up a friend like Clem and making him realize that he's got a whole lot of spadework to do before he succumbs to the farm fever. Sometimes I'm too late and cause more offense than common sense.

It is for my friends, like Clem, and many of you who have similar cases of farm fever, that I have written this book. It is intended for those of you who have decided that *now* is the time to turn that dream of "a place in the country" into a successful reality.

In the pages to come, I will try to give you some real, honest-to-goodness encouragement. I will try to point out some of the pitfalls that lie in wait for you down many picturesque country lanes. I'll offer some helpful hints and practical pointers designed to change a family of city gardeners into a family of successful farmers. I believe that you will find it a handy guide to making your dream of a good and simple life in the country come true.

Or you may decide, after reading a few more chapters, that farming is not what you thought it would be and perhaps not the best life for you and your family. In either case, I hope that this book will be of some service in helping you make the right decision for you and yours.

If you do decide to go ahead and try farming on a full- or part-time basis, you will discover that life on a farm can be both fun and profitable.

—Jerry Baker

Everyone of us is a farmer;
for the keeping of the earth
is given to the human race.

—Liberty Hyde Bailey, "Universal Service"

Section One

On Becoming a Farmer and Getting a Farm

I've been told that there are 90 million gardeners in America, and I believe it. And, if you are willing to count everyone who is trying to make friends with a plant and help it grow, there are probably 10 million or 20 million more.

Most of us get a kick out of the growing game, and our small successes breed new enthusiasms. Pretty soon, the lady who succeeded with Philo Philodendron is taking on a bigger project with Harriet Kitchen Herb and her whole family! The young father who planted a tree a few years ago when his son was born is now helping the local boy scouts turn a trash-filled vacant lot into a pocket park!

There's something about gardening that puts us back in touch with Mother Nature and the whole out-of-doors! I wonder how many of those 90 million or more American gardeners are farmers at heart? How many of us dream of one day owning a place in the country where we can once again enjoy the taste of clean air and the delightful smell of living, growing things that we've planted and cultivated ourselves? How many city or suburban moms, watering their plants, and dads, mowing their lawns, have a secret yearning to walk in their own country garden or stoop to feel the soil in their own fresh-plowed fields? Well, no one knows for sure, but I'd be willing to bet that the dream of owning a little farm is a dream that's shared by millions of folks just like you and me.

In the not-too-long-ago days that were called the "New Frontier," it was a commonplace occurrence for President and Mrs. Kennedy to entertain outstanding Americans at special White House dinners. At one of those glittering functions, when the dining room was filled with authors, artists, composers, inventors, scientists, and poets, the president arose and welcomed his guests with a toast. The White House, he said, had not been honored with an assemblage of so much genius since the nights when

Thomas Jefferson sat down to dine alone! I wonder how many of the amused and applauding guests that night stopped for an instant to reflect that our third president was, first and foremost, a *farmer!*

It's said that when Jefferson was vice president, urgent business called him one day from his beloved Monticello to Baltimore, where he was to meet with an important member of the Adams administration. Weary and dusty, still wearing his farm clothes, he rode into town and went directly to the leading hotel and asked the host for a room. After looking the "hayseed" over from head to toe, the snobbish manager told him that there were no rooms available and suggested that he try another hotel that catered to tradesmen and farmers. Thinking nothing of it, Jefferson went around to the more modest establishment where he was welcomed and given a comfortable room.

A few hours later, the government official arrived at the first hotel and asked for the vice president. And, after Mr. Jefferson was described, the insolent manager realized his mistake! He rushed to the modest inn to see the vice president and apologize. He breathlessly told Jefferson that his hotel staff was preparing the best and most luxurious quarters for his use. But Jefferson calmly told the hotelkeeper that he was quite content to stay where he was. He said that he felt more comfortable in an inn which catered to *farmers!*

Thomas Jefferson envisioned that the young United States would someday become a great agricultural nation and that the bulk of our people would make their living from farming. He felt that there were certain degrading features connected with an urban life devoted to the pursuits of commerce and industry. In 1801, when he was president, nine out of every ten Americans lived on a farm. Today, that statistic has been almost exactly reversed and experts are saying that by the end of this century, 96 percent of all our citizens will live in city homes and apartments with only about 800 square feet of living space!

The Flight from the Farms—and Back!

Millions of Americans have a farm in their past. For Ilene and me, our grandpas were both farmers. But in the past two generations, most of the members of our two families joined the mass exodus to the cities, which began after World War I and escalated after World War II. That old song, "How You Gonna Keep 'Em Down on the Farm after They've Seen Paree?," became just as true as it was popular. In the past thirty-five years, more than 30 million of us gave up our family farms and life in the country for the conveniences, comfort, and recreational offerings of our large cities. The "Baby Boom" after World War II and the enormous increase in farm mechanization added greatly to the flow of our people off the farm and into city life. The use of expensive labor-saving machinery made farming on a larger scale much more economical and, as a result, farms are getting larger every year. Farm size has jumped from an average of 170 acres in 1940 to about 350 acres in 1970. Despite a plethora of government subsidies, land banks, and farmer finance schemes, many small, single-family farms are going out of business and are being gobbled up by larger competitors.

For many of the small-time farmers, those failures mean leaving or selling acreage that had been homesteaded and held in their families for generations and moving to low-income urban housing and unskilled factory work. In addition, the need for manpower has also been decreasing in agriculture. Today, about 40 percent *more* farm products are being produced by 15 percent *less* people than were needed twenty-five years ago!

However, as the charm of city life wears on and on (if not *off*), a great many of us whose families left the farms a generation back are now discovering that Thomas Jefferson may have been right when he wrote, "I view the great cities as pestilential to the morals, the health, and the liberties of man." As this century rolls to its conclusion, millions of us have begun to give serious consideration to the *quality* of our urban lives. Lots of us have come to hate the congestion, the hectic pace, and the everyday threats to our sanity and safety which are so commonplace in our crowded surroundings. Many of us fear for the future of our loved ones and deplore the fact that we have become so completely divorced from the rhythms of nature and the touch of the earth. It gives us pause when we hear predictions of crop failure and famine around the world and we stop to realize that never in our history have so many of our citizens been so far from the production of our own food.

A recent survey by the USDA's Economic Research Department found that 40 million Americans living in and around urban centers would gladly trade their present life-style for that of a small town or farm. The majority of American farms are still *family farms* and today more and more people are eager to go back into the country and take up farming as at least a partial way of making a living. Planners at the USDA predict that family farms and farm incomes will increase dramatically in the next decade or so. They also predict that most of the farms in this country will continue to be *family-held* and that a sizable number of city people will acquire rural property.

I believe that this move back to the country and *the good life* is a good thing for America. Ours has always been a nation of gardeners and farmers. It's up to all of us "green thumbs" to keep it that way, because a people who are out of touch with their own land are out of touch with their past and with their future.

A Vote for the Country

Keep in mind that moving to a family farm and the commitment to hard work and long-term goals that it entails have to become a project subscribed to by your entire family. I'm not saying that you shouldn't be the boss, but it makes no difference if *your* heart is in the country if your wife and children have *their* hearts tugged by the trappings of a big town. If you go ahead with your plans against their will, you're in for almost certain disaster.

Now I'm not suggesting that you put your idea of living on a farm to a family vote right away. A lot of good candidates and good issues succeed on the merits of a hard-fought campaign. Use good gardening techniques to get the desired decision: implant the idea of country living gradually; cultivate it carefully; fertilize it with lots of food for thought; finally, weed out objections one by one with some solid groundwork.

One of the easiest ways to let your family develop a love for farm life is to let them see what it is like for themselves. Nothing accomplishes this better than taking a few family drives through the countryside.

I can remember when the automobile was more of a luxury and novelty than it is today. In those days, after World War II, our family would all pile in the car for a Sunday afternoon drive. There were no freeways or expressways. So, more than likely, we'd head out into the country to get out of the way of other "Sunday drivers"!

There was no air-conditioning, so we'd all roll down the windows and breathe in the fresh country air. Because we went off the main highways, each ride was an adventure—exploring the back roads to see what might be hidden around the next bend. Sometimes we'd have a picnic near a creek or under some big oaks or hickory trees. Sometimes we'd stop at an attractive farm for fresh eggs or produce. Of course, while Mom and Dad were making their purchases and looking around, we kids would be all over the barnyard taking in the sights, sounds, and smells of "rare creatures" like cows, turkeys, rabbits, horses, pigs, geese, and goats! The prospect of raising animals "of their own" like these can be a telling point when it comes to selling your children on the decision of whether or not to give up the city. But take my advice and don't try to buy their votes with the promise of a pony, or you'll probably end up taking care of it yourself! For kids from eight to eighty, a regular program of drives into the country can be a rewarding and fun family experience.

I suggest that you future farmers look carefully at the different farms. See how each one is laid out and what crops and animals are being raised. Check the exposure of the farmhouse and the fields. Look at the barns, poultry coops, sheds, and outbuildings. Most farmers are proud of their land and what they have done with it. It shouldn't take too much persuasion to get them to show you around and give you some pointers. Especially if you show a genuine interest. It's amazing how much you will be able to learn from another man's experiences, successes, and mistakes. You'll soon see that the same qualities and character traits which have made you a successful gardener are what make another man a successful farmer. Planning, pride, patience, and persistence are all necessary to achieve profits on the farm. And, since the farm is both the family home and a place of business, the successful farmer will have combined these same "Four Ps" in managing his home and family as well as his barns, fields, crops, stock, and machinery.

What to Look For

While you are "window-shopping"—looking at farms and country property—try to get a feel for the "lay of the land" on the various farms that you see. Don't hesitate to ask your new farmer friends for pointers in what to look for. Be on the lookout for farmhouses and fields that are on a rise or gentle slope and facing the south or southeast. Those houses get the morning sun and the southern breeze. My Grandma Putt used to say, "No neighbor can bring as much cheer to a body's doorstep as the sun."

Look for property that slopes very gently—with luck toward a creek or running water. That usually means the land has good drainage. Look for orchards or woodlots. These are things that take considerable capital and many years to grow successfully. They can add immensely to a farm's value.

On the other hand, gullied land that shows the signs of erosion, boggy land that gives indication of poor drainage, or stony, hilly land that shows signs of poor vegetation and infertile soil should all be on farms that you "see, but never buy!"

A farm with a view is always a plus. Remember that the new farmer's wife will need just as much inspiration from her kitchen window as the man out in the fields. Still, amenity values like "views" are things that must be measured against a lot of other, more important, factors for most new farm families who want to make a living off their land.

Other things to look for are wells, telephones, roads, and the distances the really good working farms are from the nearest towns and cities. For an experienced farmer, it's important to be near a good primary or secondary road so he can get his goods to market easily. Think about this if you have been considering farming in such out-of-the-way places as Alaska, Appalachia, or the Ozarks. Of course, the farther away from town you go, the cheaper the price of land will be. But keep in mind that these outlying farms are much more isolated. They are awfully far from the snow plows in winter and from schools, shopping, doctors, recreation, repairmen, spare parts, and, most important, from the markets for the farm's products.

Most farmers take the distance from their markets into account when they plan on the kinds of products they will raise on their farms. Closer in to the suburbs, land is more costly and it is often put to use for milk production, dairying, poultry egg production, or truck farming. These products don't necessarily require much land but they do demand quick and ready markets. Far out in the country, where land is cheapest, farmers raise grain, fruit, and grazing animals which require large areas of pasturage. In between these two extremes, farmers are often very diversified—growing some grains, some fruit, dairying for ice cream and butter, or perhaps raising beef, sheep, or hogs. Of course, in your travels, you will see many of these kinds of farms. When you study them closely, each is as different from the next as one snowflake is to another.

If your drives into the country and your campaign extolling the virtues of country living seem to be paying off, it's probably time now to form a family research team and get down to the difficult business of deciding where in the country you all think you would like to live and what kind of farming you want to do—time for some "R, D & P." I can't emphasize too strongly how important this step is in relation to the ultimate success you will have as a farm family.

Try to think of your family as a corporation about to embark on developing and producing a new product. Research and development are critical. You must set some realistic long- and short-term goals. You must decide on your products and have an educated idea of your potential market. You will need to carefully plan a system for organizing your time and labor force. In most cases, this labor force consists of you, your wife, and your kids. Finally, since there are so many natural forces and outside factors that will come into play, it's not a bad idea for the family to all join hands and ask God's blessing.

The USDA tells us that for every five families who set out to become established in farming, four fail. Your research, development, and prayer will go a long way toward making *your* family the one in five that succeeds.

The Pioneer Spirit and the Homesteading Myth

You don't have to be "far out" to be a farmer. A lot of us are steeped in the age-old tradition of determined and hardy pioneers who came to this virgin continent and carved a family farm out of the wilderness. Small wonder the idea excites us; it's heroic and historically true. But if you've never tried farming before, I'd advise you *not* to trade all your belongings for some seed and a few tools and go homesteading in Alaska, Amazonia, or Australia. While all three places would probably be glad to get you (Amazonia gives immigrant farmers 20 pesos, 20 acres, and a water buffalo), your chances of success are almost surely two—slim and none. Sure, Alaska has some homesteading land left and at least you won't have to give up your citizenship. But Mother Nature doesn't always act like a fellow American, and when she gives the cold shoulder you really know you've been snubbed! And, if your farm venture should happen to fail, you might just find out that in Alaska you and your family can be just as stranded as a New Yorker who's looking for a taxi in a rainstorm! If you still insist on becoming a trailblazer and a divorcee from the twentieth century, please prepare as thoroughly as possible for what will be in store for you. Write to Manager, Land Office, Anchorage, Alaska 99501, or Fairbanks, Alaska 99701, and have them send you all the information they can about homesteading and farming in the forty-ninth state. (For homesteading information regarding Amazonia or Australia, check with your nearest Brazilian or Australian consul.)

Before you leave, try to build up a bankroll that will provide you with backup income for at least the first three years, even if this means acquiring the land first and working at your current job until you know as much as you can about what you are getting into and you have saved enough to make success possible. It's not a bad idea to try farming closer to home first to see if the rewards of the profession live up to your expectations.

In addition to these far-out regions already mentioned, there are other homesteading possibilities. Although there is very little free public land left in the temperate zones of the United States that is suitable for farming, the federal government does (on rare occasions) free a few remaining tracts of land for homesteading through the Bureau of Reclamation and the Bureau of Land Management of the Department of the Interior. These may be homesteaded when water becomes available. Even with water, most of these land tracts are in arid and semiarid parts of the country. You will have to do a lot of expensive and back-breaking labor over several years to bring this land into a state of fertility which will allow profitable farming. Before the government will give you a homestead tract, you will be required to personally examine the land. For information on obtaining this kind of land as a homestead or as a lease or outright purchase, write to the Bureau of Land Management (or the Bureau of Reclamation), Department of the Interior, Washington, D.C. 20240.

Getting Back to Basics—Country Property

Now that I have hopefully put a damper on your pioneering spirit, let's go back to your reasons for wanting to go into the country. Let's find out what you hope to get out of it, and what, considering your present financial and family situation, you can reasonably expect to achieve *immediately* and farther down the road.

Lots of families are more *philosophically* tuned in to the idea of year-round country living than they are actually ready to break away from the System and their present situation. If you are among these, don't give up on the idea! One of the most sensible courses open to you may be to find some property that your family can use *now* for a weekend and summer home and that you can use *later* as a place for you and your wife to retire to. This means that you will be looking for a weekend vacation home or campsite that can, over a reasonable period of time, be converted into a comfortable year-round country residence.

The chances are that you won't be using this property very often for farming the land, so you may want to look into the possible profits to be made by renting it out to tenant farmers or sharecroppers, etc.

Your goal may be to join the hundreds of thousands of families who acquire rural property as a way of investing their money. They are betting that the current upward trend in rural land values will continue and see this way of tying up their money as being a hedge against inflation. Often, they just hold the property and pay the taxes without making improvements. They realize that it is the land and not what's on it that is most valuable. Most economic planners agree that inflation is growing. Various prognosticators see this growth as continuing at a rate of anywhere from 6½ to as high as 15 percent a year for the next ten to twenty years! You can see that investing in land, which is increasing in its real value at about twice the current inflationary rate, makes sound investment sense. Improving this investment land can go a long way toward insuring your chances of making a good profit with country property. Another way to increase profits is through rental income.

You may plan to rent the farmhouse on such property as a vacation home and the other buildings for storage and garage space for harvested crops or farm equipment. Rental income obtained in this way may easily offset mortgage payments, taxes, and assessments, and may even bring you some additional profit. It's often a good idea to pour this profit back into improving the land and the buildings in hopes of someday moving there yourself or doubling or tripling your original investment by waiting for the right moment to sell.

If investment and profits are your primary goals, look into crop-share leases, livestock-share leases, or tenant farmers who will pay you rent on the farmhouse, barns, buildings, and the land. If this is your intention, be sure to protect yourself so that your partners or tenants are required in their lease to keep up the property and make improvements on the buildings and especially in the fertility of the land. Otherwise, you may lose a whole lot more in the long run than you gain when the rent and share checks come in.

Another possibility for a long-term investment is to turn a good-sized parcel of the country property you intend to buy into woodland. Tree crops take many years to mature but require very little care. They hold back erosion of your soil and their leaf-mold provides a mulch of natural compost which will add to the fertility of your land. There are other advantages which you can find out about from your county agent or from USDA bulletins listed on page 264. Or write to pulp product manufacturers, like American Plywood, 1119 A St., Tacoma, Washington 98401, or the Abitibi Corp., 1400 N. Woodward Blvd., Birmingham, Michigan 48011.

Part-time and Full-time Farming

The United States Department of Agriculture defines a part-time farmer as one who lives on a farm with his family, gets some money and food from farming but gets a major part of his income from nonfarm-connected jobs.

A great many people go into this type of farming because they are repulsed by the quality of their lives in the cities and want to move to a small town or rural environment where there are social conditions and activities which they feel are more suitable for raising a family. Others become part-time farmers because they love the land and take a special pride in growing plants and raising animals.

About six years ago, Bob Timmons, a doctor friend of mine, sold his thriving city practice in Chicago and bought a more modest one in a small town in Henry County, Illinois. He and his family of seven have been avid gardeners for years, so it doesn't surprise me to learn that they are successfully operating a three-acre vegetable truck farm near Kewanee on a part-time basis. Dr. Bob says his kids are learning the rewards of hard work and developing a good sound business sense by marketing their crops themselves. They are also putting away money for their college educations, and still finding time for a full social calendar which includes: 4-H, FFA, school dances, junior American Legion baseball, the local church's young people's group, and so on. Believe me, a three-acre vegetable farm is enough to keep three or four full-grown men going full time!

Another doctor friend of mine moved from the city to a hundred-and-twenty-acre horse farm near Davidson, Michigan. He and his family love horses and now raise breeding stock for trotting. The major part of his income comes from work he does as a veterinarian, which he receives by traveling the Midwest State and County Fair Trotting Circuit during the summer months.

But don't think you have to be a doctor to be able to move into the country and take up part-time farming. My Uncle Bob has farmed a hundred and seventy acres for more than thirty years. For almost all of the period, he worked full time as an accountant in Flint. I've always thought of him as one of the best-durn farmers in the world!

Joe Komperda has fifteen acres outside of Armstrong Creek, Wisconsin. Of course, the whole world, with the exception of two grocery stores, a post office, a Roman Catholic church, and five taverns, is *outside* of Armstrong Creek! At any rate, Joe lives on fifteen acres near a lake. Ten or twelve of his acres are a woodlot. The rest consists of his home, his garden, and his lake frontage. Joe, his wife, Joan, his son, Joey, and daughter, Vicky *all* work full time in nonfarm-connected occupations. Because of his trees and garden—which raises enough fruit and vegetables for his entire family, including three more children, Nina, Richard, and Marian—Joe must be considered a successful part-time farmer. You can be one too with just a little money and a whole lot of pride, patience, and persistence! Part-time farming offers a feeling of security for retired people and others on fixed incomes who want to supplement their pension incomes with the farm and garden products they raise and sell or consume themselves.

The old saw "Don't Bite Off More Than You Can Chew!" goes back to the days when most country people chewed tobacco. That "little bit extra" that you couldn't work on in one corner of your cheek often got swallowed—and anyone who ever swallowed even a tiny bit of chewing tobacco knows what trouble really is!

So it is with folks who take on too large a part-time farm or too big a nonfarm job. If you get up early and work until you're exhausted on the farm, you won't have a lot of zip left for your city job. If all your time must be spent at the office, someone else will have to do the constant chores which are a daily part of farm routine. Emergency situations have a way of coming up on a farm just as much as they do on your nonfarm job. You can't be in both places at once, so one or the other will have to suffer. That's why I advise you to take on the work of part-time farming *gradually*. (Don't bite off more than you can chew!) My Grandma Putt used to say that a *half* well done is better than a *whole* half done. So plan to put your potentially productive land under cultivation a little at a time, year by year. And above all, don't buy a full-time farm and try to run it as a part-time farmer! If you do, you're bound to lose three things: your city job, your farm, and your health.

If you finally locate a good piece of land suitable for your planned part-time farming needs and subsequently discover that it is only offered for sale as part of larger acreage, then you have a big decision to make. If you never want to be more than someone who likes to live in the country, pass it up. If that's not the case, you might want to purchase the entire tract in preparation for the day you—or one of your children—will go into farming full time. In the meantime, it's probably best to let some of this land lie fallow under some nitrogen-producing cover crop, or "green manure." This will help hold back erosion and increase the fertility of the soil.

If you have to put this land into production in order to meet your mortgage payments, don't try to do it all by yourself while you hold a nonfarm job at the same time. Instead, I suggest you look into crop-share or stock-share or cash leases. Make sure that the contract you sign says that you provide the land and your partner provides the labor. Another alternative is to rent off some of this property to neighbors who may want to farm it.

I'd be careful about buying a big chunk of land in the hopes that you'll be able to turn around and sell off a large part of it almost immediately at a profit. Remember, farmers may not say much, but they're usually pretty good horse traders. So, unless you know a whole lot about rural real estate values in a particular locality, be extremely careful.

Actually, you can raise all the fruit and vegetables your entire family can eat—and then some—on about a half-acre of land. However much more you buy depends on how much income you need to get out of your farm activities.

If you plan to work full time, year-round, off the farm, hire out as much of the heavy labor as you can. Be practical, don't expect to work every waking minute of every day of every year. If you do, you'll be sorry. Well-organized, practical farmers take their families on vacations just like well-organized, practical city folks. Remember this, when you plan your future, "A farm is not a place to *spend* your life, it's a place to *live* it!"

The family looking for a full-time farm will have to plan for a different set of opportunities and problems. Careful study and planning are even more important for the prospective full-time farm family than for those who want to acquire country property for other purposes.

Farming full time requires: family cooperation, devotion, and hard work. It requires a knowledge of agricultural sciences, business acumen, and an abiding faith in, and love of, the land and all that grows upon it. It is a back-breaking job, fraught with an untold number of natural and man-made perils including droughts, floods, pestilence, and having the bottom fall out of the market in the crops or stock you worked all year to raise.

While intelligence and a grasp of modern farm production and management techniques are essential prerequisites for success, nothing can outweigh the importance of having common sense and confidence in your own ability to profit and prevail.

To be a full-time farmer you must be a long-range farmer. You must learn that Mother Nature runs on her own peculiar time schedule and that there is a lot of love and labor between the sowing and the harvest. You must realize that there are bound to be bad years as well as good years. You can only hope that you will learn from your mistakes and from your experiences and that your life as a farmer will leave you, your family, and the land a whole lot richer.

Finding Good Country Property

The first problem a city family that wants to return to the land will run into is that there just aren't very many decent farm properties available for sale. The USDA says there are about 3.4 million farms in the United States today. That depends on what you would be willing to classify as a farm. The U.S. Census defines farms as ". . . places of three or more acres . . . if the annual value of agricultural products, exclusive of home garden products, amounted to $150 or more."

From that definition, I'd be willing to bet that hundreds of thousands of those 3.4 million "farms" aren't exactly what you had in mind when you decided to move to the country!

About 150,000 farms become available on the market each year. But highway construction, land developers, manufacturing plants, airports, conservation projects, schools, and a whole host of other interests are taking thousands and thousands of acres of good farmland out of production. And, as I mentioned previously, mechanization has made farming on a large scale more economical, so when good farmlands do become available, they are often bought up by neighboring farmers who wish to increase their harvests or livestock range. Also, the sons and relatives of established farmers are the first to hear about farms that come on the market in a particular area. Despite all of these factors, some farms do manage to get left for "city slickers" like you and me. The trick then is for us to figure when, where, and how to go about finding the best of these—and especially the one in particular that fits into our own personal dream about country living.

When to Look for Your Farm

Knowing *when* to look for farm property can be just as important as *where* you look. Around harvest time, farm work becomes hard labor. That's when an older man begins to think about retiring. If the crops have been bad in a particular year, the older farmer may become discouraged and think about spending his later years in retirement. My Uncle Bob began to think that way just about every year in the late summer and early fall.

"You know, Jerry, after the crops and stock are sold this year, I think I'll find me a city slicker who wants to enjoy the 'easy life' here in the country. I'll sell him the farm here and move into a nice little suburban place where 'the living is tough'!"

You can see that the grass is always greener in the other guy's yard. Anyway, late fall and early winter is the best time to catch a farmer when he's tired and in a selling mood. Late winter and early spring have a way of getting his juices going again, and from planting time on through the summer your chances of getting good, producing land are pretty slim.

Where You Farm Does Make a Difference!

Where you decide to locate your farm or country property will be influenced by several overlapping determining factors. Among these are your goals, your financial situation, and what, if any, farm products you plan to produce.

Hank is a bachelor who is not very interested in working the land. He wants to invest in country property that he can use for fun, games, and profit. His property will probably be located in a scenic area near the ocean, a lake, ski area, good fishing stream, or hunting grounds. He will want property that's within a few hours' drive or handy to air or rail transportation. His investment plans may be influenced by future development that will turn nearby land into a resort area, thus potentially increasing his property values.

Mike and Alice want to retire on a fixed income. They will be looking for country property with a small but livable year-round residence that is far enough out from large cities and suburbs where they can purchase the land outright at a good price. A place where their mortgage payments won't be so high that they will cut into their savings or pension income. They will want a small place, probably just a few acres. They would like to have a small garden, chickens for home egg production, perhaps a couple of hogs or a cow, and a small orchard. They will need a small barn or enough farm buildings to house a tool shop, garden tools, and perhaps a minitractor. They will be concerned about the severity of the climate and will probably want to locate near a small town where there is good shopping, adequate health care, a church of their persuasion, an American Legion post, a women's club, and one or more other fraternal and social organizations to which they may become affiliated. They probably won't be traveling to the city very often, but will want a place that their sons and daughters and grandchildren can reach easily in one or two days' drive.

Jerry and Mildred, my mother and dad, had similar needs, but since Dad is a ham radio operator, he wanted an acre of land on a hill where he could get good radio reception. They decided to live in a large and modern mobile home, so they looked carefully to make sure the property had a good well and septic system. Their place has a small vegetable garden where they get plenty of exercise and raise much of the food for their own table. Needless to say, they love the country.

Frank and his family moved from town to operate a nursery and do some part-time truck farming on the side. They wanted no more than half a dozen tillable acres but ended up buying ten because the location fitted so many of their other requirements. The place was adjacent to a good road that is traveled by heavy traffic to the city. They needed frontage on the road so they could sell their nursery and vegetable products from a stand or simple store. Frank required a place with fertile soil, adequate drainage, and a good source of water. He plans to build a greenhouse on the frontage property. The location was close enough to the city so Frank has no problem moving his "farm" products to market and nursery retailers. He didn't need much land, but he did have to pay a premium price on his acreage because of *where* it had to be located. Be sure you consider this when you buy.

Dean worked in the southern California aerospace industry, but things got slow and he and thousands like him were laid off. Both Dean and his wife, Margie, had grown up on farms, so even though they didn't have much money, they did know how to farm. They also have four sons. Two of the boys were of high-school age and were willing workers. The family's decision to go back into farming was unanimous.

Dean and Margie had grown up in the Corn Belt where most crops grow well and dairy, hog, and sheep farming are common. They looked for a California farm in an area which was also suitable for "general," or cash-grain, crop farming.

Knowing what to look for and seeking advice from a reliable farm broker, Dean found a 140-acre farm a considerable distance from the nearest city where the per acre cost of land was relatively low. The land had not been farmed for several years. Generally it's a good practice to stay away from "bargain" farms like this, but Dean and his advisors checked the property carefully. They found that there was good drainage, no advanced erosion, and the soil was still fertile and potentially productive. This new farm family didn't have much capital to work with so they proceeded very cautiously.

Although the barns and farm buildings were in a sad state of neglect, Dean had made certain that they rested on good foundations and could eventually be repaired. He didn't spend any money on these structures at first. His many years on a farm as a boy and young man had given him the ability to take the long view of where he was going. He was determined to put his money where it would have the most immediate positive effects. Every possible dollar went into income-producing projects. Livestock, like dairy cows and poultry which began to pay as soon as they were acquired, were among his first purchases. Because they located their family farm in an area where diversified farming was the general practice, Dean and his sons were able to join a co-op and borrow much of the heavy equipment they needed. With this they began to bring the land back into production, *gradually.* (Remember, "a half well done is better than a whole half done.") By not spreading his seed, fertilizer, or labor over too much territory at once, he was able to have good results that first year. By making friends with the people in the community and listening to the advice of experienced local farmers, many years his senior, Dean and his family began to make good headway. He has now made needed repairs to the farmhouse and rebuilt the barn. Hard work and good marketing know-how, plus the help of his co-op, has enabled him to increase his equity from slightly less than $2,500 to $12,500! This amazing and continuing success story has occurred in slightly more than four years! His 140-acre farm, which he bought for $26,950, is now evaluated at $55,000, an increase of more than 100 percent! Dean's oldest son is going to a nearby state agricultural college. Margie has become a leader in the women's group of her church and this year they have a young exchange student living with them and going to school with their second son. Another son is studying in Germany as part of that same program.

Because they took the time to find the right farm in the right area and then proceeded to work hard and become upstanding and respected members of the farm community, Dean and Margie and their children have become successful full-time farmers.

Because they went into their new farm operation with careful planning, previous experience, and the willingness to listen to local experts, they were able to make "pay-as-you-go farming" pay off.

So, whether you want a place in the country for recreation, investment, retirement, to live in and raise a family, or to operate a full-time family farm, you might do well to follow the example of one or more of these people in deciding *where* to locate. The next step is to work out any financial problems *before* you take that big plunge.

As I write this book, I sense that millions of Americans are looking toward their own financial futures with a great deal of fear and uncertainty. Some of the so-called experts see the entire fabric of the world's economic system ripping apart from the crises of world famines caused by African droughts and Asian monsoons. Even here in America, increasing costs of producing essential raw materials, including farm products, seem to have the "little guy" trapped in a dangerous funnel cloud of ever-upward, spiraling inflation.

No wonder so many of us have decided to hedge our bets and leave the cities where we just can't seem to keep up. It just seems to make good sense to reinvest our lives, money, and labor in the land and the production of food.

As inflation pushes farm product prices up, it also thrusts up the costs of buying farm acreage, housing, equipment, and other operating materials. One of your major problems, then, will be to get the most out of your capital and labor investment. If you keep a positive mental attitude and use your common sense, you will be able to do this even in the uncertain times ahead.

What Can You Really Afford to Buy?

Maybe you have visions of yourself as a country gentleman on a large-landed estate. Or maybe you see yourself as starting out to the country with little more than the desire to work and the shirt on your back. Most likely, though, a careful examination of your real assets and liabilities will show that your true financial situation lies somewhere between those two extremes. To find out for certain, fill out the "Family Financial Evaluation Sheet" on pages 24–26. It will help you figure out your true net worth and how much capital you have available for that farm in your future, right now. This work sheet is meant to be a tool not a test. So be honest with yourself as you fill it out. Be sure to include the income of every working member of your "family farm corporation."

FAMILY FINANCIAL EVALUATION SHEET

Family income (*present annual income and expenses*)

Personal income

A. Husband ______________

B. Wife ______________

C. Other ______________

D. Other ______________

Pensions ______________

Annuities ______________

Miscellaneous ______________

TOTAL ANNUAL FAMILY INCOME ______________

Expenses

Housing

A. Loan, taxes, insurance ______________

B. Rent ______________

C. Utilities	________
D. Phone	________
E. House maint. and repair	________
F. Household help	________
Food, beverages, tobacco	________
Clothing	________
Personal allowances	________
Transportation	
A. Autos, loans, and insurance	________
B. Auto repair	________
C. Public trans. or other	________
Medical, dental, and personal care insurance	________
A. Life	________
B. Health and accident	________
C. Personal property	________
Recreation and entertainment	________
Education	________
Church and charitable donations	________
Gifts	________
Personal loans	________
Furniture, goods and equip. purchases	________
Miscellaneous	________
Personal business expenses	________
A. Husband	________
B. Wife	________
C. Other	________
D. Other	________

FAMILY FINANCIAL EVALUATION SHEET

Personal income taxes
- A. Federal ______________
  ____
  ____
  ____
- B. State ______________

TOTAL ANNUAL EXPENSES ______________

PRESENT SURPLUS INCOME
(*income less expenses*) ______________

Now that you know your net worth and what you have in the way of investment capital, you can begin to take the first steps toward becoming established in the country. Plain old common sense will tell you that "moving to the country" and "getting established on a country place" are two completely different things.

Whatever your goals and reasons for acquiring country property, you know it's going to cost you money. If you are ever going to attain your goals and enjoy country living the way you want, you must operate in a systematic and businesslike manner right from the beginning.

A good way to start is by separating your personal finances from your "country place" finances. You should do this even if your country place is mainly to be used as recreation, retirement, or investment property. Just go down to your bank and open a "country business" checking account. I would call mine the Jerry Baker Country Company. Make sure you get a business-type checkbook with adequate space to keep accurate records of all transactions.

At the same time you open your checking account, it's a good idea to open a savings account in the name of the same company. You might not need it right now, but it will come more and more into play during the next few years. Even at this stage of the game, it will help you keep your farm and personal money separated. Now transfer to these accounts all the funds you are willing to earmark from your personal net worth *surplus* for your country place project. At first, you will probably be concerned at the small amount of investment capital you have to work with, but at least you will know where you stand and this will help you decide how quickly you can take the next step. From now on, all transactions connected with your investment in country living should be made through this account. It should never again be allowed to mingle with your personal finances until such time as you decide to sell or abandon your country living "company."

How Much Money Will You Need?

There's an old saying in the Baker family, "Money isn't everything, but it sure helps!" I'm sure we Bakers don't have the corner on that choice bit of knowledge, but keep in mind that a careful assessment of the money you have and the money you will need to have are all-important before you go out shopping for your country property or farm.

Now that you know how much money you have at the present time, let's turn our attention to one more important piece of information. *How much money will you need to purchase your country property?* The answer to this question may be highly speculative since it depends on how drastically you and your family are willing to change your current standard of living to achieve your country living goals.

Most of us who live in the city or suburbs have come to enjoy a style of living that allows for a good many little "extras" which make us a little more comfortable than the average family around us. For many people, these "extras" are what make life worth living. However, if you have just discovered that you don't have a whole lot of surplus money to invest in order to get your country living project underway, maybe you'd better sit the whole family down and begin reevaluating your priorities. Now is the time to decide if your family has to maintain its current standard of living or if you are all willing to sacrifice a little comfort in order to "get where you want to go."

Don't kid yourself by thinking that it will cost you a lot less to live in a rural setting. Actually, recent comparisons have shown that the cost of living isn't all that lower down on the farm. The little bit that you might save here and there in one part of your budget will balance with the extra expenditures you will have to make for other items.

If your evaluation of your family expenses and net worth shows that you are living beyond your means, obviously your dream of owning a place in the country will have to be delayed until you can readjust your standard of living to a lower level and build up some surplus capital to invest in country property. This may mean that you will have to maintain your present job or get another better-paying job in order to maintain your current life-style and to increase your country savings. You may be able to do this by selling your home in the city, buying a less expensive home in the country, and commuting to work. By farming part-time, or renting off your fields to neighbor farmers for a share in the crop or stock they produce on it, you may be able to gradually make a go of it.

Whatever course you decide on, the first step toward accomplishing the job of raising country capital is to put your family on a good tight budget. Be practical and create a budget which allows your family to live well within their "comfort zone" and at the same time lets you save toward your goals. Otherwise, you may all become quickly discouraged and ignore your best-laid plan. (Many good books which tell you how to set up such a budget are available at any public library or bookstore.) Keeping to such a budget, there are very few of us who can't reduce our total family expenses by a figure between 10 and 20 percent.

As you make out your budget, try to decide which family members will be keeping their present jobs or getting new ones, and which ones will be working full or part time for your farm company. Try to make an honest projection of what you and your family can reasonably expect to make during the time frame you have set as a deadline for getting established on the farm.

Let's suppose you have $15,000 which you can transfer to your "country company" account and that in the next five years you would like to be established in the country in a place, including the farmhouse, that is worth around $150,000. You are thirty-five years old and want to be well established by the time you are forty. That is, in five years you want to be making enough money from your country property to maintain your family at a standard of living which will fulfill their requirements of health, education, comfort, and welfare. You want to be your own boss in running the farm operation. You want to feel secure enough about your tenure on the land to know that you hold it through at least the next growing season and harvest. You also want to have a controlling share in the ownership of all crops and other farm inventory necessary for operating at a profit. You have set a goal for yourself that is obtainable if you plan your cash expenditures carefully and implement your plan gradually. Don't try to buy everything at once, or you may run out of cash at a critical time and have a major setback.

City people often think that the costs in purchasing a farm will be similar to the kinds of costs they incurred when they bought their home in town. Before you plunk down all your hard-earned cash on the first attractive farm that appears to be within your price range, stop to consider what your cash outlay priorities will be.

If you are buying a country place for recreation, you probably can't go too far wrong if you buy a well-built old farmhouse with a view that has woods for hunting and a well-stocked stream or pond. It may not matter to you that the land is rocky and eroded, or that the barn and the outbuildings are in a sad state of disrepair. Perhaps by maintaining a good job in a nearby city you can come up with enough cash to make any needed repairs to the house and to keep your new place in the country looking attractive.

Remember though, if you plan on living on the place full time without putting the land into any kind of production, you may still be faced with a lot of hidden costs. Before you buy, try to determine if the well is adequate to give you and your family a reliable supply of water. Check the plumbing and septic system, the heating and electricity to be sure they are all in good working order. Make sure you are also buying enough good working equipment to keep the weeds from overrunning the farmyard and the snow from blocking your access to main roads in the winter. Try to hold on to as much cash in your "country company" account as you possibly can to be available when those hidden upkeep, taxes, and assessment costs arise. You may be shocked when you find out how much money is required to keep even the most modest country place looking decent!

Part-time farming can be a pretty inexpensive operation. Anyone who has ever had a successful suburban garden knows how easy it is to produce more than your family could ever use, even on a little 10-by-20-foot plot. If you are planning on raising only enough food to supply you and your family with vegetables, fruit, eggs, and milk, you probably won't need much more than an acre of tillable land, some hand-gardening tools, seed, fertilizer, and a small cowshed, chicken coop, and tool room or garage. You also must figure your own time and labor as one of your costs. An hour or two a day should be sufficient, on this kind of an operation, to feed a family of four or more. This will give you a surplus of food to can, freeze, sell, or give away.

Use Your Family Labor Force Wisely

When you and your family become farmers, you should also become do-it-yourselfers whenever possible. Remember, your family labor force represents money. Spend it wisely on work that

APPROXIMATE AMOUNT OF LABOR NEEDED EACH MONTH FOR SELECTED HOME-PRODUCTION ENTERPRISES IN THE EAST CENTRAL STATES

Enterprise	Hours of labor needed during—												
	Jan.	Feb.	Mar.	Apr.	May	June	July	Aug.	Sept.	Oct.	Nov.	Dec.	Total
Garden, 1 acre, well-diversified, prepared with horse or tractor power			30	110	110	50	50	50	50	50			500
Field corn, 10 acres, cut, shocked, husked by hand			15	20	35	25	8		30	65	32	10	240
Hay, 10 acres							10		70	40			120
Milk cow, 1 *	20	20	20	20	20	20	10	10	25	20	20	20	225
Laying hens, 12	6	6	6	6	6	6	6	6	6	6	6	6	72
Pigs, 2				8	8	8	8	8	8				48
Bees, 1 or 2 colonies				2	1	2	2	8	2	3			20
Rabbits, 1 buck and 4 does	10	10	10	10	10	10	10	10	10	10	10	10	120
Milk goats, 2	15	14	14	13	12	12	10	10	10	20	20	15	165

* Shift in time of freshening would shift monthly hours of labor in various months.

will bring a high rate of return for the sweat expended! USDA Farmers' Bulletin #2178 on the preceding page includes a chart which will help you figure the approximate amount of labor you will need to put into selected home-production enterprises. While this chart was developed from the experiences of part-time farmers in the east central states, it can be a helpful guide wherever you buy a farm.

Avoid the Initial Cost of an Overlarge Spread

Don't make the mistake that other new farmers often do by thinking that all you have to do to increase your production and make more money is to buy a lot of land. It's often very expensive for a part-time farmer to increase production. When you enlarge the size of your farm operation, figure on the increased cost per acre of the additional land. This cost of course depends on local land prices, but you have no guarantee that they will be the same or less than what you had to pay per acre on the original tract you purchased. They could be much higher if the additional land is extremely fertile or in a choice location. Also figure on needing more money for tools, machinery, seed, fertilizer, livestock, farm buildings, feed, outside labor, etc.

If you plan to raise crops or stock on a full-time basis, you will have a somewhat different set of cash-outlay priorities from those of the family who come to the country for recreation or part-time farming.

Obviously, for crop farming, the single most important thing for you to put your money into is the land. Next comes seed and fertilizer and water. Next, the farm machinery and equipment to plant, cultivate, and harvest your crops. You will need storage facilities and some sort of conveyance to bring your crops to market or distribution centers. You may be able to rent many of these items, but then rental costs should be figured in to your cash-outlay priorities. The farmhouse, barns, stock, and other items of farm inventory may rank farther down on the list from these key uses for your initial investment of capital.

You can write to your state agricultural extension service to obtain figures on sample costs involved in almost any type of farm operation. Costs for specific areas can be obtained by contacting farm advisors or county agents.

The State of California Agricultural Extension Service bulletin on "Sample Costs to Produce Crops" by A. D. Reed gives the new crop farmer a very good idea of the per acre cash-outlay costs he can anticipate in raising various farm crop commodities. The bulletin compares these costs to yield per acre figures obtainable under good management. It also includes a list of definitions of the costs involved which you may find helpful:

Labor cost—includes social security, compensation insurance, housing, transportation, and other labor costs in addition to the wage paid the laborer.

Fuel and repairs—the cost of fuel, oil, and lubrication plus the total cost of repairs (labor, parts, etc.).

Material—includes seed, fertilizer, livestock feed, water or power, spray, machine work hired, and other costs not included under labor, fuel, and repairs.

Equipment overhead—depreciation, interest, and taxes.

Harvest—the total cost of harvest, usually up to the point where the farmer is paid for the product.

Cash overhead—office, accounting, legal, interest on operating capital, and other costs of management.

Rent—includes the rent actually paid for the land where available or the cost of taxes, interest on investment, and depreciation on fixed facilities which the operator incurs by owning the land.

Management—usually calculated at five percent of the gross income.

While the cost per acre of growing everything from alfalfa to wheat in California may be of no immediate interest to you, similar cost breakdowns for the type of farming you intend to go into can be obtained from the extension services in your area.

After analyzing these costs versus the probable yield per acre, the size of the farm you are considering and the equipment, machinery, and supplies which will be included in the purchase, you can get a pretty good idea of how much of a business can be bought with the funds you have available. It's important to remember that *you are buying a business*. Don't put yourself into a situation where you can't win or where you have to go so far into debt that it will take you years to put your farm business on a profitable basis. An unexpected drought or rainy season could cause you to lose everything if you get too far behind in your mortgage payments.

First-Year Cash Outlays

Experienced farmers will tell you that there are several ways to shift or minimize your first-year cash requirements. You can eliminate buying costly new machinery by hiring out many of the jobs that can be done with this machinery to custom operators. Or you can rent or borrow the machinery and do the work yourself. Some farmers purchase used machinery and then rent it out to neighbors when they are not using it. If you do this, make sure your agreements with your neighbors include their picking up a good share of the additional maintenance and depreciation costs you will incur. Often tractors, plows, disks, planters, combines, mowers, cornpickers, balers, wagons, and manure spreaders are owned jointly by co-ops. As a member of the co-op, you will have the use of this jointly owned equipment on a rotation basis.

If you intend to specialize in some sort of stock farming, you can cut down on your initial cash outlays and first-year expenses by not purchasing bulls, stallions, boars, etc. Instead you can shift the cost of maintaining these animals into other necessary areas such as feed, housing, fencing, and sanitation by taking advantage of low-cost artificial insemination services when available. Where no such services exist on a professional level, you may be able to obtain them from a neighboring farmer for a minimal stud fee.

Feed costs are another high-priority cash requirement for farmers with feeder or stock-raising operations. Many stockmen complain that these costs are becoming so high that they can barely make a profit. While there is little you can do about these costs during your first year of operation, you can plan ahead to defray some of these costs in future years by raising your own feed. If you have tillable land available on your stock farm, look into the business sense of growing your own feed. If not, investigate the cost of renting nearby acreage. Of course, the costs of planting, raising, harvesting—and storing—this feed must compare favorably with projected costs for purchasing such feed elsewhere. These costs must also be included in your annual cost-return balance sheet.

Quality Versus Quantity

Buying good quality stock is another important consideration. For example, a few milk cows that are good producers are often a smarter buy than many cows which give less milk. A few meat animals which bring the highest dollar at the market are a better buy than many animals which bring fewer dollars per weight.

Animal quality can be extremely important in other ways. Auburn University researchers found in a recent study of a feeder pig operations that returns on investment were very favorably affected by acquiring high-producing sows.

"The producers who sold more than 15 pigs per sow had significantly lower fixed costs per pig than producers who sold less than 15 pigs per sow. The producers who sold 15 or more pigs per sow earned 17.9% on their average investment. The producers who sold less than 15 pigs per sow earned *a minus 13.2% on the average investment.*" * [Italics are author's]

From these figures, you can see how important it is to take quality into account when you are investing your farm capital.

Holding Some Cash in Reserve

When you purchase your farm you may be tempted to put all your cash into high-priority cash-outlay expenses. By doing this you hope to cut the amount of your farm mortgage payments to their absolute minimum. While this may seem to make good business sense, it is just as important to hold some cash back to meet unexpected contingencies. These cash reserves can sometimes *save* you a great deal of money. For example, one of the many problems connected with raising crops like corn, soybeans, wheat, or rye is that you will have to wait a whole growing season after you put your cash into planting before you can expect to get any return on your money. You may have to store your grain and wait even longer if the prices at harvest time aren't high enough to make it worth your while to sell early. After planting, you may need cash to invest in other farm operations like poultry and dairying which offer an immediate return on your investment dollars. After harvest, it's true you may be able to take advantage of low-interest federal loan programs like the one that will give you $1.39 for each stored bushel of harvested wheat, but in the meantime you may need cash to solve any number of unexpected farm problems. Similar emergencies have a way of occurring in all

* "Costs and Returns of Producing Feeder Pigs in Alabama," Bulletin #407, Agricultural Experiment Station, Auburn University, November 1970.

other types of farming. So again, I advise you to be extremely careful about your initial cash outlays and to hold some of your cash in reserve to cover contingencies.

In summation, in order to get off to a good start with your farm, treat it like a business. Invest your money as gradually as possible and according to plan. Try to get your money producing a good return as quickly as possible so that you have something to sell. Hold some cash in reserve to meet any temporary money crisis that might arise. Use as much of your early profits from farm sales to increase your farm resources and inventory. Be a good money manager and you will probably turn out to be a good farmer.

Section Two

Finding the Farm for You

Caveat Emptor!

"Hey, Rube, here comes another City Slicker!" is a common joke in some of the farm communities that surround our major metropolitan centers. You'd be surprised how many people who have learned to survive in the city think they know a whole lot more than their country cousins. Don't be misled by slow drawls and down-home colloquialisms! There are just as many shrewd country shysters, looking out for innocent city people whose inexperience is all too apparent, as you will ever find in any town. Also, many old "horse-tradin' " farmers aren't above selling their rundown farm and eroded untillable fields to some poor unsuspecting dude. Finding the right farm for you will require all the good "horse sense" you can muster. Appearances can be deceiving, so go about your farm finding in a careful, businesslike way. Don't lose your cool and jump at the first country "castle" you find with a "contented cow" munching weeds in the front forty. Despite what the owner or real estate agent tells you, the "castle" may be crumbling inside and the "contented cow" may be a Trojan horse full of tricks and trouble.

It always amazes me when a well-educated and successful city dweller, who would dispassionately and carefully check out any business proposition in town, can't wait to throw his hard-earned money down the first rathole he finds in the country! Would you buy a store in the city or suburbs without checking to see if it is productive and worth the money? Of course not!

If you do become interested in a place you see in the country and find you are being pressured into buying it, back off and try to find out why. But don't ask the realtor or the seller who are pressuring you. Instead, check with the neighbors on either side. Check out the fertility of the land by having the county farm agent take a look at it and have a soil test made. Check with the folks at the local Grange, United Farm Organization (UFO), or community newspaper. The editor will probably tell you how long and how often the property has been advertised. Ask the minister or priest if he's heard any negative things about the farm you intend to buy. Check the county recorders' office to find out when the place was last sold and for how much.

Now you may think this is an awful lot of embarrassing bother. But it's not half as bad as having a few or all of those people telling you the "bad news" *after* you buy! If everything checks out and the land is good, you will have earned a good reputation as being careful and conservative in your business dealings—and that's not so bad!

Where to Get Good Information and Sound Advice

When you are looking for a farm or piece of country property, begin by trying to take full advantage of every good piece of information and sound bit of advice that you can. Some of your best sources will be: the United States Department of Agriculture farmers' bulletins, state agriculture extension services, county farm agents or farm advisors, reliable farm realtors, farmers' newspapers and journals, professional farmers' organizations like the Grange and United Farm Organization, local farmers, and others in a given farm community. You can also get a lot of good books and publications on every aspect of agriculture from your public library.

First, look at some maps. Try to decide what state you want to farm in or what kind of farming you want to do. (By 1910, almost all of the land in this country had been put into production by knowledgeable farmers, so today it's fairly easy to find out what has traditionally grown best, where.) You will see that the best yield per acre of almost every region has been carefully recorded. Only intensive injections of fertilizers, time-consuming labor, capital, and *water* can turn the semiarid and nonfarming areas into productive agricultural areas. If you don't have the time, money, or water, don't disregard the experiences of farmers who have tried and failed to make an area or a piece of property that is best suited for one type of farming operation profitable with another.

Climate and Geography

Just a quick look at the maps will show you that certain crops and certain stock animals are raised most often in specific parts of the country. Don't get the mistaken idea that some wise-guy mapmaker down at the USDA just drew those regions on there to make a pretty design like some of us used to do back in fourth grade geography class! Actually, a great deal of care is put into assessing all the climatic, topographic, watershed, and farm production data to be translated onto the maps. The climate in a given geographic area is what will let you know what kind of farm you can locate there. Agricultural production depends primarily on these climatic factors:

1. temperature (maximum/minimum range within which plants will grow)
2. latitude and longitude
3. altitude or elevation
4. length of growing season
5. light
6. prevailing wind conditions
7. availability of water

Generally, the closer to the equator, the higher the temperatures, the longer the growing season, and the more light available. Then, each degree of latitude away from the equator means the length of the growing season is shortened, the amount of sunlight is lessened, and the maximum/minimum temperature range within which individual crops grow is decreased. Finally, as we approach the Arctic Circle, nothing grows on the tundra except certain mosses and lichens. Above the Arctic Circle, vegetation ceases to grow entirely.

The effect of geography on what you can grow on your farm is enormous. It has been estimated that for each degree of latitude away from the equator, flowering is retarded as much as four days; for each five degrees of longitude from the East Coast toward the West, flowering is advanced five days; and for each 400 feet of altitude, flowering is retarded four days.*

Altitude or Elevation

One of the great sightseeing attractions in Colorado is the cog railroad which carries visitors to the top of Pike's Peak. As I traveled up the steep incline one day last autumn with Ilene and my friends Joe and Lorraine Daley, I couldn't help but notice the changes in temperature and vegetation as we slowly climbed toward the 14,000-foot summit. Every few minutes, we passed into a new environment. Each successive environment supported a different type of plant life. From 6,000 feet up, we left the birches and the aspen behind and entered into the domain of the conifers. Then, suddenly, at 13,000 feet we passed the timberline and it was almost as though we were traveling toward the Arctic Circle. From then on the altitude caused the temperature to drop below the point that would support any shrubs or grasses. Here or there a marmot would stray upward for some tasty sedge or moss. Finally we reached the peak, completely covered with a pack of ice and snow.

The mountain is a good illustration of the maximum/minimum range of temperatures that can be found over most of our northern hemisphere. If, for example, the base of the mountain was at sea level, the traveler would pass through every major vegetative environment between the equator and the Arctic Circle on his trip to the top.

Crops and the native grasses which support livestock have

* M. S. Kipps, *Production of Field Crops,* McGraw-Hill, 1970, pg. 32.

naturally adapted to the various climatic and environmental factors where they are raised. Certain plants, like Irish potatoes, are called cool-season plants because they are tolerant of the cool weather, excessive precipitation and short growing seasons of states like Maine and Idaho, which are in the higher latitudes. Sweet potatoes, on the other hand, need hot days, warm nights, and a long summer growing season. They grow best below the Ohio River or in the irrigated lands of the West and Southwest. Similarly, cotton grows best in the South where the growing season is two hundred days or more, but oats which need cooler weather, plenty of precipitation, and a short growing season do best in the North. Tomatoes, a warm-season crop, are a little more tricky. These warm-season plants require warm days, 75° or above, but need cooler nights, down around 65°, in order to set their fruit. That is one reason why southern California and the southern shore of New Jersey, two areas which traditionally cool off at night, are almost always among the leaders in tomato production.

Length of Growing Season

The length of the growing season is usually figured from the date of the last hard frost in early spring to the date of the first hard frost in late autumn. Below is a USDA list of hard frost dates for various parts of the United States which you may want to use as a guideline.

HARD-FROST DATES
from USDA weather records

State	First in Fall	Last in Spring
Alabama, N.W.	Oct. 30	Mar. 25
Alabama, S.E.	Nov. 15	Mar. 8
Arizona, No.	Oct. 19	Apr. 23
Arizona, So.	Dec. 1	Mar. 1

HARD-FROST DATES
from USDA weather records

State	First in Fall	Last in Spring
Arkansas, No.	Oct. 23	Apr. 7
Arkansas, So.	Nov. 3	Mar. 25
California		
Imperial Valley	Dec. 15	Jan. 25
Interior Valley	Nov. 15	Mar. 1
Southern Coast	Dec. 15	Jan. 15
Central Coast	Dec. 1	Feb. 25
Mountain Sections	Sept. 1	Apr. 25
Colorado, W.	Sept. 18	May 25
Colorado, N.E.	Sept. 27	May 11
Colorado, S.E.	Oct. 15	May 1
Connecticut	Oct. 20	Apr. 25
Delaware	Oct. 25	Apr. 15
District of Columbia	Oct. 23	Apr. 11
Florida, No.	Dec. 5	Feb. 25
Florida, Cen.	Dec. 28	Feb. 11
Florida, South of Lake Okeechobee	almost frost-free	
Georgia, No.	Nov. 1	Apr. 1
Georgia, So.	Nov. 15	Mar. 15
Idaho	Sept. 22	May 21
Illinois, No.	Oct. 8	May 1
Illinois, So.	Oct. 20	Apr. 15
Indiana, No.	Oct. 8	May 1
Indiana, So.	Oct. 20	Apr. 15

State	First in Fall	Last in Spring
Iowa, No.	Oct. 2	May 1
Iowa, So.	Oct. 9	Apr. 15
Kansas	Oct. 15	Apr. 20
Kentucky	Oct. 20	Apr. 15
Louisiana, No.	Nov. 10	Mar. 13
Louisiana, So.	Nov. 20	Feb. 20
Maine	Sept. 25	May 25
Maryland	Oct. 20	Apr. 19
Massachusetts	Oct. 25	Apr. 25
Michigan, Upper Penn.	Sept. 15	May 25
Michigan, No.	Sept. 25	May 17
Michigan, So.	Oct. 8	May 10
Minnesota, No.	Sept. 15	May 25
Minnesota, So.	Oct. 1	May 11
Mississippi, No.	Oct. 30	Mar. 25
Mississippi, So.	Nov. 15	Mar. 15
Missouri	Oct. 20	Apr. 20
Montana	Sept. 22	May 21
Nebraska, W.	Oct. 4	May 11
Nebraska, E.	Oct. 15	Apr. 15
Nevada, W.	Sept. 22	May 19
Nevada, E.	Sept. 14	June 1
New Hampshire	Sept. 25	May 23
New Jersey	Oct. 25	Apr. 20
New Mexico, No.	Oct. 17	Apr. 23
New Mexico, So.	Nov. 1	Apr. 1
New York, W.	Oct. 8	May 10
New York, E.	Oct. 15	May 1

HARD-FROST DATES
from USDA weather records

State	First in Fall	Last in Spring
New York, No.	Oct. 1	May 15
N. Carolina, W.	Oct. 25	Apr. 15
N. Carolina, E.	Nov. 1	Apr. 8
N. Dakota	Sept. 13	May 21
N. Dakota, E.	Sept. 20	May 16
Ohio, No.	Oct. 15	May 6
Ohio, So.	Oct. 20	Apr. 20
Oklahoma	Nov. 2	Apr. 2
Oregon, W.	Oct. 25	Apr. 17
Oregon, E.	Sept. 22	June 4
Pennsylvania, W.	Oct. 10	Apr. 20
Pennsylvania, Cen.	Oct. 15	May 1
Pennsylvania, E.	Oct. 15	Apr. 17
Rhode Island	Oct. 25	Apr. 25
S. Carolina, N. W.	Nov. 8	Apr. 1
S. Carolina, S. E.	Nov. 15	Mar. 15
S. Dakota	Sept. 25	May 15
Tennessee	Oct. 25	Apr. 10
Texas, N.W.	Nov. 1	Apr. 15
Texas, N.E.	Nov. 10	Mar. 21
Texas, So.	Dec. 15	Feb. 10
Utah	Oct. 19	Apr. 26

State	First in Fall	Last in Spring
Vermont	Sept. 25	May 23
Virginia, No.	Oct. 25	Apr. 15
Virginia, So.	Oct. 30	Apr. 10
Washington, W.	Nov. 15	Apr. 10
Washington, E.	Oct. 1	May 15
W. Virginia, W.	Oct. 15	May 1
W. Virginia, E.	Oct. 1	May 15
Wisconsin, No.	Sept. 25	May 17
Wisconsin, So.	Oct. 10	May 1
Wyoming, W.	Aug. 20	June 20
Wyoming, E.	Sept. 20	May 21

The amount of sunlight available in a specific geographic region or location is another factor you should consider when you look for a farm location. All plants need light in order for their seeds to germinate and for the mature specimens to carry on *photosynthesis*. This complex process is the way in which they manufacture their own food from the nutrients they extract from the soil.

If you should happen to purchase a farm in a shady meadow or mountain valley that receives sun for only part of the day, you will soon discover that light also affects crops in another way, called *photo-periodism*. The term photo-periodism refers to the length of day (or hours of sunlight) a plant needs in order to flower. Some plants require *long-days* (approximately eight to fourteen hours of sunlight); others require *short-days* (approximately four to eight hours of sunlight). A complete list of the long- and short-day crops which will grow in a given location may be obtained through the county agent or state extension service.

As a farmer, you may not be as much interested in bringing your crops to flower as you are in some other aspect of their growth. It may be more important to produce strong, healthy vegetative growth. For example, corn—a short-day crop—needs long periods of sunlight in order to reach its full profit potential. In other words, this is one of those things like farsightedness and nearsightedness. The short-day crops—like corn, sorghum, and soybeans—need long periods of sunlight each day during the growing season. The long-day plants—like oats, red clover, timothy, and wheat—can be best brought to harvest in areas where there is a shorter growing season with days that have relatively few hours of sunlight.

Often the climatic factors which determine good production and high yields for various crops can be overcome or changed for the better if you plant the proper type of seed. The USDA and many major companies have spent vast sums in research in order to develop new hybrid varieties of crop seed which are designed to get the best results in specific growing areas. If you are concerned about the potential yields you can expect in a particular farm location, check before you buy. The USDA, state agricultural extension service, or one of the many field representatives of the major seed companies will be happy to answer your questions. Remember, you are a potential customer for the seedman, so he will be delighted to make suggestions regarding the best crops to raise on the farm you are looking at.

Prevailing Winds

When you look for a farm, be sure to take into account the prevailing condition of the winds. Such winds can have a beneficial effect on your crops by carrying moisture to your fields from nearby bodies of water. Or they can create a drying effect if they should happen to blow your way from a desert or dry plain. If the latter condition exists, make certain the farm you are interested in has plenty of water available to counteract this constant negative situation.

Water

Locating on the windward side of water can also increase the length of the growing season, as water temperature changes slowly and tends to affect the surrounding countryside by making it less subject to sudden frosts.

Too much, or too little, water available for crops and livestock can also be a big determining factor when it comes to deciding on a piece of farm property. Damp, boggy land may be fine for rice or cranberries, but it may be just as impractical as dry, semiarid land for the farmer who wants to raise less specialized crops. If springs and streams on the property only have water in them during or right after the rainy season, you should not be tempted to count on them to bring water to your crops or livestock.

If you are looking at land that is broken up into several pieces, or the fields are separated by roads or other barriers, make sure you can get water to the stock or crops in each area. This may mean digging several wells. Where this is not possible due to lack of water availability or because of high costs, the land may not be much good for farming. Trucking water is rarely an economical proposition except on an emergency basis. Even if it were, trucking water to several sites can become a full-time job.

Obviously, a dependable water supply is of paramount importance to your farm and family needs. This supply can come from wells, ponds, streams, lakes, pipelines, or irrigation sources.

If there is a pond on the property, check it carefully. It may be only a surface pond which fills in the springtime but which gradually dries up during the summer when you will most need the water. These surface ponds can become foul and stagnant. They are sometimes a breeding area for mosquitoes or pests which carry animal diseases.

If a pond is well planned and deep, it will add immensely to the value of the property. The federal government, through the Soil Conservation Service, often shares the cost of constructing ponds with farmers and stocks them with fish. However, sometimes this cost-sharing comes with strings attached. In return for the government's help, the farmer may have agreed to let the public have access to the pond and its shoreline during part of the year for recreational and fishing purposes. Perhaps you will not want to buy a farm with this kind of accessibility by strangers.

Drainage

It's just as important to have good, natural drainage as it is to have a dependable year-round water supply. Land that stays wet is usually extremely acid and unsuitable for growing most field crops. Even after it drys, the crops that grow there are often shallow rooters and no good commercially. The time to find out whether prospective farmland has good natural drainage is during, and just after, a rainy season.

What's Being Farmed and Where

The largest farming area in the country is found in the southern states, stretching from Virginia into Oklahoma on the north and from Georgia to east Texas on the south, historically called the Cotton Belt. In it you will find more farmers and family farms than anywhere else in the United States. But a city family in search of farm property might be surprised to find that a quiet revolution has been going on here for years. While cotton still is "king," its production is generally moving in two directions—westward and onto huge commercial and highly mechanized farms. Since World War II, increasing sophistication in machinery has begun to do away with the "stoop labor" in the cotton fields. In their place, machines operated by a few skilled workers plant and harvest this important crop and airplanes do much of the fertilizing, weeding, and pest control operations.

The long growing season has made the region ideal for other

crops beside cotton. No longer are the southern farmers caught up in a one-crop economy. Now you can find plenty of farms which raise a variety of farm products including corn, dairying, livestock and pasture, grain sorghums, sweet potatoes, tobacco, peanuts, and wheat.

Probably the best-known agricultural region in the country is the Corn Belt. Here you will find some of the most beautiful and well-cared-for farms in the world. The cost of farms and farm acreage is extremely high compared to other areas, but the land is bountiful and the prices you will get for your farm products are much higher than anywhere else in the United States. The term Corn Belt may also be too general, as in the Cotton Belt. Although 75 percent of the nation's corn is raised in this area stretching from southern Michigan into southern South Dakota on the north and from central Ohio to northern Kansas on the south, livestock production and diversified farming are also very high. In the past few years, soybeans have become a major cash crop along with hay, wheat, and oats. Dairy production is also high in the region, making it, overall, one of the richest family farming centers in the entire world. About 40 percent of the world's corn is raised here.

Dairy production is more spread out. Dairy farming can be found almost anywhere near large cities or densely populated areas where there are ready markets. New England, New York, Pennsylvania, parts of Ohio, and New Jersey; the Great Lakes states of Michigan, northern Illinois, Wisconsin, and Minnesota; California and parts of Oregon and Washington.

Because the major emphasis is on milk, butter, and cheese production for the cities, sanitation requirements are strict and the prospective dairy farmer will have a heavy investment in equipment and facilities. Because many dairy operations are located fairly close to urban centers, taxes, assessments, and the high price of land are important factors for you to consider. Farm incomes in the dairy region are pretty much in line with the national average.

Farming and farmland in the northern and central Great Plains and the Columbia River basin in the Northwest is an expensive and large-scale proposition which relies mainly on the production of wheat. The tough economics of farming in the Wheat Belt are such that farmers must invest heavily in land and machinery (or the rental of same) in order to bring in a good crop. They must also have a sophisticated knowledge of up-to-date marketing procedures in order to maintain and maximize profits.

For example, during February of 1975, the nation's wheat farmers were receiving $5.52 a bushel for harvested wheat. By May of that year, the going price for harvested wheat had fallen to $3.52 a bushel. In an attempt to reverse that downward trend, many farmers decided to hold their portion of the record harvest of 1974 off the market. Instead of selling their wheat, they stored it, taking advantage of low-interest government loans to cover their operating expenses until prices stabilized. As you can readily see, profitable farming in the Wheat Belt, where farm operating costs and land costs are nearly double the national average, is an occupation that requires experience and know-how. It is rare when a newcomer has such inside knowledge.

Another highly specialized type of ranching that requires great experience and skill in order to maximize profits and avoid being strangled in the cost-price squeeze is raising livestock for meat. The great western range includes most of the private and public grazing lands of fifteen of our western states, plus parts of Kansas and Oklahoma. About half of the nation's sheep and a sixth of our beef cattle are raised here mostly in cow-calf or ewe-lamb operations. While becoming a cattleman isn't the easiest thing you'll ever do, it's not impossible by any means. My friend, actor Eddie Ryder, is successfully running a large string of beef in the beautiful West Bend country of Texas. I suspect Eddie always wanted to be a cowboy when he grew up! Like farming, ranching is a way of life that appeals to people who like outdoor living, hard work, and independence.

Fruit, truck, and special crops are cultivated in small areas of highly intensive farming that can be found in states from Maine to California, along the eastern shores of the Great Lakes, in Florida, and the Gulf states, the Ozarks, the southern tip of east Texas, and in many scattered pockets in the Rocky Mountain and Pacific Northwest states. These widely separated areas seem to have the continuing special climate conditions and soil characteristics necessary for raising fruit, nuts, vegetables, and special crops like sugar beets, grass, and rice.

Maybe you'd like to try raising cane? It has been estimated that there are more than nine hundred different types of farms in the United States and that no two farms are exactly alike. You and your family may have specific interest or experience in a certain type of farming—say, raising sugarcane. If you do, I suggest you write to the state agricultural extension service where you plan to locate and get as much information and as many specific state or county maps as they may happen to have available. These maps can give you more accurate and specific climatic conditions which will relate to your prospective farm's requirements.

Take a Quick Trip Around the Country

Now that you've looked at some maps and done some homework, you probably have a pretty good picture of the major kinds of farming being done in the six or seven broad farming areas of the United States. You have also picked up a few tips on the ways that climate and geography can help or hamper your efforts in agriculture.

If you haven't yet decided exactly where you want to live or farm, it might be worth your while to take a quick trip around the country so you can check some of the farms that are on the market and what their sellers are asking for them. Right now you're probably saying, "Hold it! I can't afford to go running around the country looking at farms. That would cost a fortune!"

The trip I'm talking about is free, and you won't have to leave the comfort of your favorite easy chair. You can take an enjoyable and informative tour of the rural property market just by browsing through one or more of the various farm and country real estate catalogues that are available free of charge. These catalogs give brief, thumbnail descriptions of land, homes, farms, recreational, and business offerings from Maine to Hawaii. Two of the best of these nationwide realtors who furnish free catalogues are: United Farm Agency, 1104-B Pacific Mutual Bldg., 523 W. 4th St., Los Angeles, Calif. 90014, and Strout, P.O. Box 690C, Arcadia, Calif. 91006. Other sources for such catalogues can be obtained from your local Sunday newspaper or real estate board. The catalogues make great dream books, but be extremely careful about *purchasing* properties by mail. All reliable realtors will insist that you carefully inspect country property *in person* before you make any purchase commitments.

Your County Agent

If you already have a pretty good idea of where you want to locate, take a day or more to become acquainted with the county farm agent or farm advisor. He will know right away if the type of farm you are looking for is available in his county. He may even know of a farm or two that isn't listed but which you might be able to persuade the current owner to sell. Of course he's not in the business to do this and you will have to use your own good judgment as to whether you can even ask him for such a favor. At the very least, he will be able to tell you if your farming plans sound reasonable and if you ask him, he will probably be able to suggest several reliable local realtors or farm brokers who will help you.

The Man Who Sells You Your Dream

When the time comes that you know you're really ready and serious about buying country property, then it's time to get a good, reliable realtor. Remember, the man who sells you your farm or rural acres is rarely the man who owns it. Most likely he will be someone whose business is selling real estate.

Finding a good real estate broker or agent who really works hard on your behalf is like finding a good family doctor who will make house calls! If possible, I'd advise you to find a realtor—a real estate broker who has passed the qualification exams for active membership in a real estate board affiliated with the National Association of Real Estate Boards. These men are almost always extremely qualified and usually licensed by the state or local government. The term *realtor* is not a synonym for *real estate broker* or *real estate salesman* or *real estate agent.* A *realtor* is a person who is a member of the National Association of Realtors and one of the local real estate boards affiliated with the National Association. This professional organization has a code of ethics which its members are asked to sign and which may be even more restraining and specific than the local real estate law and the local Laws of Agency. Realtors are expected to conduct themselves competently, reflect the highest integrity, and be fair-dealing in their business affairs.

Some real estate brokers or agents who have not been accepted into the Association of Realtors are still fine, upstanding businessmen. It may be that there is not a great demand for buying and selling real estate in their area and so not enough agents operate within the area to constitute a board. It may be that a local businessman sells real estate in his spare time and therefore feels it's not necessary to join the local board.

Unfortunately, there are some salesmen or brokers who see the profession of selling real estate as a good way to make a fast buck. They may work for a big company which is developing a tract of land, or they may operate on their own. These guys usually know a whole lot about what is within the letter of the law, but very often they ignore the dictates of common ethics and the "golden rule." Usually a little checking up on your part will let you know what kind of a man you are dealing with. Buying any type of land can be tricky, so try to avoid doing business with a tricky salesman.

Whether you select a realtor, a real estate broker, or a real estate salesman, you are protected under the Law of Agency in the local or state civil code.

The Law of Agency

The Law of Agency governs the activities of *agents* and *principals.* Simply speaking, an agent is someone who contracts or agrees to act for another person, or principal. Most of us come in contact with agents when we deal with attorneys, insurance salesmen, stockbrokers, and real estate brokers and salesmen.

The principal is usually a seller or a buyer who contracts the agent to act for him. Technically, an agent is more than an employee. He has been placed in a position of confidence and trust and has many more responsibilities than an ordinary employee. He is bound, by the law, to disclose all the facts relating to a particular transaction and the handling of his principal's property. The agent is not permitted to use his principal's property to his own advantage. And perhaps most importantly, the agent may not gain any monetary interest in the property of his principal without first obtaining the principal's consent.

You should keep in mind that the real estate agent normally represents the seller's interests more than yours, as he is most often the seller's agent. However, the law usually requires that he honestly represents the principal's property to prospective buyers like yourself. He is obligated to fully disclose to you any information he receives from the seller or elsewhere which is pertinent to the sale.

In other words, let's suppose the broker learns there are nematodes infesting the soil of the farm you are looking at and that this problem will make farming the land next to impossible. Under the law, he is bound to disclose this information to you even though it will ruin his chances of making a sale. However, if he does tell you about this problem, and you ignore it or don't bother to investigate, then the agent is relieved of any responsibility of misrepresentation, and you will have no recourse under the law. So, hire a competent professional and be sure to follow through on his advice and information.

Sometimes the seller holds back information pertinent to the sale of a piece of property or home from both the prospective buyer *and the broker*. Under the Law of Agency, he too is obligated to the concept of full disclosure. For example, a seller may not tell you that the farmhouse has a faulty septic system. You buy the farm and move in, then you have trouble and call the local repairman, who tells you that he told the previous owner that this septic system would have to be replaced. Now, you decide that you're not going to take this lying down! You try to sue the seller and the broker because they both told you that there were no defects in the plumbing or septic system. When the facts come out in court, you may find that the broker misrepresented the worth of the property to you through a mistake or because the seller had not disclosed the negative information to him. The judge may charge the seller with damages or rescind the sale, but he will probably not hold the broker responsible.

Do Business with a Specialist

One way to eliminate many of the problems like those we've been talking about is for you to find a broker who is a specialist in selling the type of property you are looking for. Most of us have only dealt with real estate brokers who sell residential property—the fellow who sold you your home in the city or suburbs. However, there are brokers who limit their business to selling commercial property, investment property, farm property, retirement property, etc. You might do well to find a farm broker or a man whose business it is to know good property of the kind you are seeking.

Usually, when you buy a home in the city you never see the real estate agent again after closing or your first day of occupancy. The nice thing about buying country property from a well-established small town realtor or farm broker is that you're bound to see him around from time to time. A good realtor knows this. He knows that if he does a good job for you, you will sing his praises to your city friends or relatives who share your dream of "a little place in the country." He also knows that the seller is usually on his way *out* of the community and that unless he has family ties to bring him back, the broker will probably never see him again. And since the seller is leaving the place, he probably won't recommend that his friends move there.

Hiring Your Own Broker

You may not know this, because it's not the usual way of acquiring a country home or farm, but it is perfectly possible for you to hire your own broker and make him your exclusive agent in seeking a rural location. What you do is sign a contract with a broker giving him a fee for seeking farm property for you. As *your* agent he will make it his business to find just the kind of farm you are looking for. As your agent he will seek out farm properties that may not even be listed. He will try to get you the best value for your money. He will not be so interested in seeing that the buyer gets top dollar. He will tell the seller that he represents you and that the commission will come from you. This may make the seller more willing to deal. It may influence the seller to take a much lower offer than he would under normal circumstances because he knows a professional is bidding for you, someone who won't easily be influenced by appearances.

By hiring your own agent, it just might be that you will pay the same amount of money in the end, but you will be insuring that the things that ought to be checked and rechecked for the buyer are taken care of.

Whether you decide to hire your own agent or select one who is the agent for the seller, here are some helpful hints on finding the right realtor and getting the most out of him.

Get in touch with a realtor who's been recommended by someone you know and whose judgment you trust. Make an appointment and tell him what you have in mind to buy. Be very candid with him about your financial resources and about the kind of farm you would like. The more he knows, the more he will be able to serve your best interests.

If possible, on this first meeting, get him to take you out and show you some of his listings. This will not only give you an opportunity to get a feel for local land prices and what's in the area that's available, but it will also show you what kind of farming is being done in the area. By paying close attention to what your successful neighbors are doing, you may be able to pick up some valuable information which will come in handy when you commence your farm operation. You will probably also see some mistakes that local farmers are making. It's always good to learn from the other guy's mistakes whenever you can.

This first outing will also allow you to size up your man and see if he seems to be a straight-shooter, as far as you are concerned, or if he's merely anxious to close a deal—any deal.

Take into account that he's probably sizing you up, too, to see if you're serious about buying or just wasting his time. This first tour of the local farm community will give him an opportunity to see if you really know what you want, and what you want to spend. As I mentioned in Section One, many people haven't taken the time to evaluate their financial situation. Most people see themselves on a millionaire's farm which they bought for ten or twenty thousand dollars. He will probably show you properties listed at much more and much less than you told him you could afford to pay. He does this to let you find out for yourself if you've been dreaming. This first excursion will help you come down to earth and do some serious figuring, if you haven't done so already.

If you are as well prepared as I hope you are, you really should not need this little lesson of his, but go along for the ride anyway—you just might stumble onto a bargain.

If you happen to see something you like, don't snap at it. You can always go back for a second and a third look. Sit down with paper and pencil and write down a list of the things you want to check on your return trips.

If you don't see any property that arouses your interest, maybe it's your fault. Perhaps you didn't give the broker enough details regarding your requirements. If not, do it now before you and he waste a lot of time looking for the wrong place. It is entirely possible that no such property exists in the immediate area. It could also be that although there are many properties that resemble your needs, none of them happens to be on the market right now. If not, the broker may want to do a little homework before you meet next time, to see if any owners of unlisted properties in the area which fit your requirements might be persuaded to sell. Or he may want to make some calls to check out leads in nearby communities or neighboring counties.

Your broker may suggest that he place some ads, on your behalf, in some of the community newspapers in the region, or perhaps in one or more of the farm journals. After you check the copy, it's probably a good idea to let him handle such advertisements because he will be able to weed out unsuitable properties easily and thus avoid your having to waste your time. While all this is going on, you and your family can continue touring and exploring the farm communities that seem most appealing to you.

Back in the time of Ilene's Grandfather Richmond, there were few, if any, real estate agents. If a man wanted to buy a farm or a piece of country property, he put an advertisement in the country newspapers and waited for replies. When he got several letters describing available farm properties, he wrote to the ones whose farms sounded the most interesting and set up appointments. Then, he hopped in his buggy and went looking. Sometimes it took many months to find something close to what he wanted. Many city people who were lured "back to the land" discovered, after they put out their hard-earned money, that they had purchased a pig in a poke.

It's ironic that today many descendants of those fall-guys of yesteryear are still willing to go to any lengths to follow in their forefathers' footsteps. If you are one of those who wants to avoid paying the realtor's commission by finding and buying *your* farm *yourself*, good luck!

It's the wise greenhorn who knows how to use the eyes, brains, and experience of a shrewd horse trader. Most good rural realtors and farm brokers have trained themselves to know good value. I recommend that you take advantage of your realtor's knowledge of the area and of the land in it. He will know soils, drainage, conservation, and all the nuances of sound farming technique. Work on his personality. Make him your friend. It won't be difficult. Most professional people are flattered when a stranger respects their competence and experience. Show this respect and I'm willing to bet that your realtor will go all out. You might get lucky and find a fine farm on your own, but the chances are against you. You will save much more than the realtor's commission if he steers you to the place you really want and then shows you the good points and the flaws. He'll take you into the country with your eyes open, and that's worth more than you'll ever pay him.

You've Got to Know the Territory

My realtor friend, Lt. Col. Willie Williams of Wrightwood, California, tells me that whenever you look at a farm or piece of rural property you must ask yourself this all-important question:

"How good is the *location* and how well does the location tie in with my *purpose* or reason for wanting to buy?"

That is not an easy question to answer as there are usually many mitigating factors involved and because it rarely happens that any individual piece of property will satisfy *all* your requirements. One way to find out about any location is to check out a wide circle of territory around it. Poverty is a big problem in rural communities. You may not want to locate in an economically distressed area. As you drive around, try to evaluate the economic stability of the farm community and nearby towns. Are the neighboring farms prosperous-looking? Or do they have a "borderline" rundown appearance? Are the main roads in the county well paved and maintained? What is the condition of the primary and secondary roads which front or lead to the farm you are considering? How good and how convenient will these roads be in transporting your farm products to market? Is there good transportation to shopping? To schools? Are the schools modern or antiquated and uncared for? Is the nearest town thriving or dying? What are the *growth patterns* of the area?

If you are buying land for investment purposes, you and your realtor should be aware of the patterns of growth of the community. Investors will want to buy a piece of property that is in the path of an expanding commercial or suburban population. If your realtor says that future development is headed for a particular location, ask him for supportive evidence. He will probably be able to show you some old maps which, when compared to recent maps, show how the town or suburbs of the city are moving along certain paths toward the location. He may be able to supply you with information of planned commercial or recreational development nearby which will boost land values. He may be able to show how new superhighways or commuter railroads are going to parallel the location.

Of course, if your purpose in coming to the country is to get away from noise and congestion, you may not want to locate where you will soon be surrounded or where the roar of trains or traffic will disturb your family. You may not have the capital to cope with the increased taxes and assessments that are sure to come with the burgeoning population.

If you are planning to raise poultry for egg production, you will not want to locate where train or traffic noise will make your hens nervous and unable to lay. So whatever your purpose in terms of buying this location, you will be smart to check out the patterns of growth.

Your Telephone Can Be a Time-saving Tool

Let's assume that by now your realtor has found a piece of country property that he feels approximates what you are looking for and he wants you to drive out and take a look. Before you gather up the wife and kids and go racing off to the country, take the time to ask some pertinent questions:

1. What is the location?
2. What is the price?
3. What down payment will you need?
4. What is the size of the property?
5. What use is the land being put to now?
6. Are any livestock, crops, farm equipment to be included in the sale?
7. What kind of farm buildings are on the location and what's their condition?
8. Does the present owner live on the property? If not, why not?
9. What kind of house is there and how old is it?
10. What are the taxes?

Since *location* is your number one consideration, have your realtor describe it thoroughly in terms of soils, drainage, availability of water, utilities, transportation, patterns of growth, etc. If you are familiar with the area, you will probably know almost immediately whether or not it meets your requirements.

Most of us have only so much money available to buy our dream place in the country, so the price and the amount of money required for a down payment are probably key questions which will help you decide whether to take the time and trouble to go look at the property.

The other criteria (questions) can be graded in importance according to what your *purpose* is in acquiring the property at that location. If you intend to farm there, you will have to find out if a successful farm business is being carried on now. If you intend to raise crops or livestock, you'll need to know about the current production, the soil, drainage, utilities, equipment, etc.

If you intend to retire there, you'll want to know about the size of the property, the farmhouse and its condition, the availability of water and utilities, and perhaps if it will be possible to farm part time or rent off part of the land to neighboring farmers. A couple looking for retirement property will also want to know about the growth patterns and how they will affect taxes and assessments in the years to come.

Using this list of ten questions, you should be able to find out enough about the property on the phone to let you decide if you are interested enough to go and see.

Be Prepared!

When you finally do decide to go out and look at a piece of country property with an idea to buy, you'll find out that it's hard to take in everything all at once. Why not follow the suggestion proposed in the old boy scout motto and Be Prepared? To do this, all you have to do is make up a checklist.

Most state extension services or state agricultural colleges will either have such a checklist or they'll be able to tell you where to get one. Perhaps your realtor or county agent will have one that's been designed for the particular farm area. Here are a few of the things you will want on your list:

TYPICAL FARM SCORECARD

Type of farm: ______________________

I. Size and amount of productive land
   A. Acreage: ________ # Fields: ________
   B. Acres in woodland or timber: ______________
      1. Recreational ______________
      2. Commercial timber ______________
      3. Woodlot ______________
      4. Orchard crops ____________ Types and value ____________
   C. Acres in crop production ______________
      1. Single-crop farm? ______________
         Type of crop and acreage ______________
      2. Multi-crop farm? ______________
         Rotation? ______________
      3. Types and acreage in production
         a. ______________ d. ______________
         b. ______________ e. ______________
         c. ______________ f. ______________
   D. Acres in pasture ______________
   E. Acres in roads and lanes __________ Condition __________
   F. Acres in swamps and bogs ______________
   G. Acres in streams, ponds, or lakes ______________
   H. Acres irrigated ______________
   I. Acres suitable for crops if irrigation added ______________
      Cost ______________
   J. Acres not suitable for production of any kind __________
   K. Estimate of total crop value ______________

II. Acres or area in buildings (Total) ______________
   A. Farmhouse __________ Age __________ Condition __________
      1. # Stories __________ Exterior Area __________ x __________
      2. Roof __________ Age __________ Condition __________
      3. Foundation ______________
      4. Heat: Elec. ______________ Gas ______________
         Nat'l. ______________ Propane ______________
         Tank owner __________ Type of heater __________

TYPICAL FARM SCORECARD

5. Water heater: Gas ____________ Elec. ____________
Size ____________ Age ____________

6. Utilities: Elec. ____________ Gas ____________
Telephone ____________

7. Garbage disposal ____________
Dishwasher ____________

8. Range: Gas ____________ Elec. ____________
Refrig.: Gas ____________ Elec. ____________

9. Washer ____________
Dryer: Gas ____________ Elec. ____________
Plumbed for washer and dryer ____________

10. Well ____________ Age ____________
Depth ____________
Windmill ____________ Condition ____________

11. City water ____________ City sewage ____________

12. Septic tank ____________ Cesspool ____________
When pumped last ____________ Loc. ____________

13. Water rate ____________ Paid until? ____________

14. Interior area: Up ____________ Down ____________
Bedrooms ____________ Baths ____________

15. LR ______ x ______ DR ______ x ______
KIT ______ x ______ SHOWER ______
O/T ____________

16. M/BR ______ x ______ BDRM ______ x ______
BDRM ______ x ______ BDRM ______ x ______

17. Furnished ____________
Unfurnished ____________

18. # Fireplaces ____________
Insulation ____________ Type ____________

19. Porches ____________ Front ______ x ______
Rear ______ x ______ Side ______ x ______
Floor: Wood ____________ Slab ____________
W/W Carpet ____________ Tile ____________

20. Cable TV ____________ Ant. ____________

21. Lawn: Frt. ______x______ Side ______x______
    Rear ______x______
22. Trees: # ____________ Types ____________
23. Driveway ____________ Paved ____________
24. Road or street paved? ____________
    Traffic: Lt. ________ Med. ________ Hvy. ________
25. Dist. to schools ____________ Town ____________
    Shopping ____________ Doc. ____________
    Hosp. ____________
26. Estimate of total home value ____________

B. Garage: Type ____________ Condition ____________
   Size ______x______
C. Barn: Type ____________ Size ______x______
   Condition ____________ Roof cond. ____________
   Elec. to barn ____________ Plumbing ____________
D. Other: ____________ Size ______x______
   Age ____________ Condition ____________
E. Other: ____________ Size ______x______
   Age ____________ Condition ____________
F. Other: ____________ Size ______x______
   Age ____________ Condition ____________
G. Estimate of total value of farm buildings excluding home ____________

III. Machinery and Equipment

A. Tractors ____________
   Age ____________ Condition ____________
B. Plow ____________
   Age ____________ Condition ____________
C. Disc ____________
   Age ____________ Condition ____________
D. Baler ____________
   Age ____________ Condition ____________
E. Combine ____________
   Age ____________ Condition ____________
F. Cornpicker ____________

TYPICAL FARM SCORECARD

Age ____________ Condition ____________

G. Manure spreader ____________

Age ____________ Condition ____________

H. Other ____________

Age ____________ Condition ____________

Est. total value ____________

IV. Livestock

A. Beef cattle ____________

Head ____________ Bulls ____________

B. Dairy cattle ____________

Head ____________ Bulls ____________

C. Pigs ____________ Boar ____________

D. Poultry: # Laying hens ____________

E. Horses: # Mares ____________ Stallions ____________

Geldings ____________

F. Sheep ____________

G. Other ____________

H. Other ____________

I. Other ____________

J. Estimated total value of stock ____________

V. Soil and drainage

A. Type of soils ____________ ____________ ____________

B. How deep? ____________ Productive ____________

C. Well drained? ____________

D. Tiled? ____________

E. Contoured? ____________

F. Suitable for diversified cropping? ____________

G. Badly in need of fertilizers? ____________

H. Best land use capabilities ____________

I. Erosion control? ____________ Needed? ____________

J. Estimated total land value ____________

VI. Seed and silage

A. Seed type ____________ Amount ____________

B. Seed type ____________ Amount ____________

C. Seed type ______ Amount ______
D. Feed type ______ Amount ______
E. Feed type ______ Amount ______
F. Estimated value of seed ______

VII. Taxes ______
VIII. Subsurface rights ______
IX. Amenity values ______

______

X. Farm income past three years ______

______ ______

XI. Anticipated farm income ______
XII. The need for machinery
A. ______ Cost ______
B. ______ Cost ______
C. ______ Cost ______
D. ______ Cost ______

XIII. The need for adequate farm buildings
A. ______ Cost ______
B. ______ Cost ______

XIV. The need for adequate fencing
A. ______ Cost ______
B. ______ Cost ______
C. ______ Cost ______
D. ______ Cost ______

XV. Seller's asking price ______
A. Down payment ______
B. Re-fi needed ______
C. Monthly payment ______
D. Interest ______
E. Present mortgage assumable? ______

Don't Let Anyone Kid You!

What land is worth to one man may be much more or much less than it is worth to another. Suppose that on a 40-acre Arkansas farm not far from Pulaski, farmer Ken Johnson finds that his land is not too fertile, so he decides to raise goats. He buys shares in a co-op which sells goat milk to hospitals in nearby Little Rock. Over a period of years, he builds up a successful operation and for the past two or three years it has grossed him between $80,000 and $85,000.

Now suppose that for one reason or another Ken decides to sell and move his goats and equipment to a larger farm in a nearby county. He says the farm is worth $75,000 and that he has been making much more on it for the past several years.

You come along and look at the farm and find out what the asking price is. You have not planned on raising goats. Even if you had, no shares are available in the co-op and the seller is taking his shares, his goats, and his equipment with him. You can't reasonably expect the same kind of farm income that he's been getting from this land. It will probably cost you a considerable amount of money, in addition to what you have to pay for the land, just to switch over into another type of farm operation. Chances are, the land has a value much lower than the price he is asking. If you can't convince him of this, you had better pass.

Don't let a seller kid you about the value of a particular farm. His experiences on the land have to be carefully weighed with an objective appraisal of the physical inventory, the worth of the land, and your plans for using it.

So You Want to Buy the King Ranch!

Many apartment dwellers head for the country with a picture in their mind of wide open spaces—of woods, streams, rolling meadows, and large fenced fields with cattle and sheep spread across them as far as the eye can see. But when it comes right down to it, they find out that they have no truly accurate idea of *size* when they have to buy farm property. They are used to measuring areas in terms of city blocks and when their realtor tells them he has found a 5-, 40-, or 160-acre farm, the city man has no idea of what those sizes are and how they relate to the various types of farm production.

To begin to get an understanding of size and area in country terms, you will have to become familiar with terms like: squares, townships, ranges, sections, quarter-sections, quarter-quarters, furlongs, chains, rods, and acres.

I've been told that out in the Golden West, when the Spaniards first settled California, they were given huge tracts of land as grants from the Spanish king or the viceroy of Mexico. A man would map out his *rancho* within the land grant by starting a rough survey from a prominent spot in the area. At sunup he would hop on his horse and ride as far and fast as he could in one direction until sundown. At sundown he'd either find a tree, rock, or stream in the vicinity or he would build a marker. The next morning at sunup, he'd leave the marker, and turning at right angles from his original direction, he'd ride hell-bent-for-leather in the new direction until sundown. In four days he arrived back at his starting point, a tired but happy *caballero*. His land was now "surveyed" and could be registered with the government.

Now that may seem like an impractical way to establish property lines, but it's actually not much different than the way the early farms were laid out in the original thirteen colonies and the bulk of New England, the mid-Atlantic, and southern coastal states. This method of surveying and establishing title to property is known as the *metes and bounds* system. Not being on horseback, the colonial surveyors would walk the land to establish a man's boundaries. Starting from an original point of reference, perhaps a stream, road, or barn, they proceeded in a straight line to the next prominent marker, perhaps a tree or rock. They would continue this way until they arrived back at their starting point. Today, many eastern farms are still legally described in this manner. The system is still used to mark the legal boundaries of irregular pieces of land.

If all the farms in the country had been surveyed and legally identified by the metes and bounds system, our country would probably be in a heck of a legal mess right now. Our courts would probably be filled with bickering neighbors, each contesting the other guy's property lines. Luckily, in the late eighteenth century, the federal government saw the flaws in the old system and sent surveying teams into the unsettled states and territories to divide up the land according to a rectangular system. Once you have an understanding of how this rectangular system works, you will be able to get a sense of the size of almost any piece of farm property anywhere in the states using the system.

How the Rectangular Survey System Works

As a government surveying team goes into an area, it begins the survey by drawing a true north-south line called a *meridian.* *

---

* Because the earth is round, all meridians above the equator end in a point at the North Pole. This makes it necessary to allow for adjustments. Therefore, the north side of each township is nearly 50 feet shorter than the south side. To adjust the rectangularity of the survey, every fifth row of townships is 35 miles square. Shortages are placed in the north and the west rows of sections of a township.

They cross this line with a true east-west line called a *baseline*. Starting from the point where the two lines cross, they lay the land out in *squares* of approximately 576 square miles (24 miles on each side). These large squares are next divided into smaller squares of approximately 36 square miles (6 miles per side) called *townships*. The townships are further divided into 36 squares called *sections*. Every section is a mile on each side and approximately 640 acres. A *quarter-section* is approximately 160 acres. Up until the early part of the twentieth century, a man could claim 160 acres, or a quarter-section, of free public land. That is why up until recent years, the majority of family farms were 160 acres or less. Today, the size of a farm is more tailored to the type of production carried out on it.

TYPICAL COUNTRY MEASUREMENTS

rod	16½ feet
chain	66 feet (4 rods)
furlong	660 feet (40 rods or 10 chains)
mile	5,280 feet (320 rods = 80 chains = 8 furlongs)
acre	43,560 sq. ft. (160 sq. rods = 1 sq. chain)
quarter-quarter	40 acres
quarter-section	160 acres
section	1 square mile (640 acres)

The Size of Your Farm

For nearly three decades now, the average size of American farms has been on the increase. There are a number of reasons for this trend, but the two most important are: 1) that the cost of farm labor has risen dramatically, forcing operators to rely on improved labor-saving equipment and 2) because this labor-saving equipment works most economically and efficiently on larger farms. There is no question that one man, with modern equipment, can do double the farm work that his father was able to do forty or fifty years ago. But if the modern farmer is going to get good use out of the equipment, it just stands to reason that he will be forced to increase the amount of acreage he has under production. It seems certain that the trend toward larger and larger farms will continue as more and more labor-saving equipment becomes available to farmers and as the cost of farm labor continues to rise.

Now if you don't plan to be part of the mainstream of American agribusiness, maybe you won't need to become a large landowner. After all, nearly half of all the farms in this country are still under 100 acres. Perhaps a smaller farm fits in better with your goals and aspirations? Also, there's no doubt that what you can afford to pay will have a great deal to do with the size of farm you are able to buy!

Obviously, you will need more land if you decide to locate in a farm region where the soils give a low yield per acre, as in the wheatland of the Great Plains states. You will need less land where there is a high yield per acre, as in the fertile farm areas of California, the Corn Belt, some of the eastern Great Lakes, and mid-Atlantic states.

But perhaps the most important consideration is *how much land will I need to do the type of farming I want to do?* Let's take a quick look at the various types of farm production and see if we can come up with any general information concerning the amount of land you will need for each.

Privacy and a Farm Garden

If your main reason for coming to the country is to regain a sense of place and belonging, a few acres should do just fine. If you don't plan on becoming a commercial farmer, even part time, size is pretty much whatever suits your fancy. Maybe a couple of gentle hills, a stand of woods, and a brook or stream? Remember that as soon as you add livestock, you will have to take their need for living space into account. Each head of cattle will need 5 to 7 acres for grazing, depending on the type of pasturage available.

You can reduce their land requirements by increasing "bought-off-the-farm" feed, but that will be expensive.

There is no doubt that a farm garden can lower your cost of living dramatically whether you carry on any other type of farm operation or not. I've already mentioned that a family of four can raise all the fruit and vegetables they will be able to eat in a year's time on about a quarter- to a half-acre of land. By selecting your crops carefully, cultivating intensively with a two- or three-area rotation system, you can get higher yields on an even smaller total garden area. Eliminate the crops that take up the most room, like corn, cukes, and potatoes.

Eight to ten fruit trees of various types will provide plenty of fruit for you and your friends if you just give them proper food, water, and maintenance.

General Farming

Most of the farms you find when you begin looking for your own special place in the country will combine the raising of livestock and the growing of mixed crops. This general type of farming is the most commonly practiced throughout the United States. It's especially prevalent in the north-central and northeastern parts of the country.

In order for you to run an operation like this, you will need to purchase land in an area where there is good summer pasture and cropland that is well suited for raising leguminous hay. While the average United States commercial farm is about 340 acres, the size of your livestock and mixed crop farm may be much smaller. In the Corn Belt, farmers have been able to increase the size of their business by adding acreage and by getting record harvests per acre. These statistics follow the improvements in farm mechanization and the increased use of this improved equipment. The size of your farm will depend partly on the type and number of animals you expect to have and partly on the methods of cropping best suited to the topography of the land. Obviously, on farms where the land is hilly and rocky, heavy labor-saving equipment will be of little value. In these areas, general farms will probably remain rather small, or they will be converted into more specialized operations which are better suited to their physical characteristics.

The larger your farm, the more you will be able to diversify efficiently. Operators who make a successful go on small acreage generally rely on one major enterprise to supply two-thirds of the total farm income. If you plan on turning small acreage into a commercial farm, you should not put too many eggs in one basket.

Fruit Farming

The cost per acre of growing fruit is relatively high. However, like truck crops and other intensive farming operations, fruit promises a good return on your investment. It's not too long ago that the Department of Agriculture estimated that one family could easily handle from 5 to 40 acres of fruit depending on the crop. The introduction of new maintenance and harvesting equipment will let you set your sights on larger-sized orchards.

Truck Farming

If you plan to go into truck farming part time or full time, you will have high costs for land, fertilizer, and labor. Even if you plan that your own family will do all the hand work or stoop labor required, this can be translated into costs—as it keeps family members from other more lucrative work off the farm.

Although I have known many farmers who worked larger areas, I feel that three acres is a lot of land for a family truck farm, especially if you don't hire extra hand labor. Three acres is almost too big for one man who doesn't work at it full time, unless he uses new planting and harvesting methods (like the mechanical tomato or onion picker) or the trash-mulch cultivation system described in a later section. A 5-acre truck farm takes about three men, working full time, and you may need extra hired help for harvesting unless you are extensively mechanized.

If you are a beginner, start small and increase your production *gradually*. This doesn't mean you shouldn't plan ahead. If the land is available, and the price is right, buy it or option it now and farm it when you have the wherewithal to devote your time, capital, and labor to obtain efficient yields. Many truck farmers who operate close in to our large cities rent all or part of their land. This allows them to free up more capital for seed, fertilizer, and other cultural necessities. When they want to enlarge their operations, they simply rent more land. Just remember that every time you increase the amount of land under cultivation, you automatically increase the risks of losses from pests and diseases. Like fruit farmers who operate on a large scale, truck farmers with big operations usually have to devote as much as a quarter of their capital to spraying, dusting, or other control programs.

The size of the truck farm you buy may be limited by your location requirements. Land for truck farming needs to be located close enough to the city to allow you to attract customers or transport your goods quickly to market. Acreage in close proximity to urban population centers is usually expensive and subject to heavy taxes and assessments. These limiting economic factors will have some bearing on the amount of land you can afford. Make sure you carefully check out all the "C, C & Rs" (conditions, covenants, and restrictions) as there may be problems regarding how and what you can farm on this land.

A truck farm is a high-cost operation. You will need money for seed, planting, fertilizer, its application, weed control, pest control, harvesting, and marketing your products. Don't invest so much in land that you have to skimp on these other vital parts of your truck farming operation.

Also, if you plan to market your fruits and vegetables yourself at a roadside stand, you will probably find that location is much more important than the size of your operation. If you locate your operation on a road that is heavily traveled, you may find yourself spending more hours *selling* than *raising* vegetables. If this is the case, you can probably increase the size of your business more easily not by trying to acquire more acreage for farming, but by handling the products of other farmers who are not so nicely located (for a fee).

If you plan on buying a fruit farm, orange grove, apple or cherry orchard that is already producing, the size of your farm will be predetermined by what is available. If you are planning to start from scratch, size will be limited by the lay of the land and cost of the land in relation to spraying, grafting, pruning, and other hand labor and equipment costs. If the land in the area where you are looking has an irregular topography, you must be selective regarding the size of the fruit farm you buy: low pockets generally are the first to collect frost and will be worthless to the fruit grower. You may have to buy more land than you need for your fruit operation because of the long time it takes to become established. For instance, small fruit like strawberries, raspberries, and blackberries take at least two years before they produce abundantly. Peaches take four years. Citrus crops take around six years. Apples take anywhere from three to twelve years, depending on the type of tree and variety. You may not be able to wait all this time before your farm begins to bring in money, so perhaps you will want to purchase extra land so you can combine your fruit operation with dairying or poultry raising or egg production.

In recent years there have been many new advances in the fruit business which are altering the size of groves and orchards. New methods of dwarfing fruit trees make it possible to use more trees per acre. For example, a standard apple tree grows about 20 to 25 feet high and 25 to 35 feet wide at the tallest and widest part of its foliage. You can plant only about thirty to forty standards per acre. Their large size increases your costs of maintenance and harvesting. A spur-type semidwarf tree grows only 12 to 15 feet high and 12 to 15 feet wide. You can plant two hundred of these on an acre! Double-dwarf trees are the latest advance. They grow only 6 to 10 feet tall, set fruit much younger than standards, and bear faithfully. Maintenance and harvesting costs are lower because the smaller trees are easier to work with.

Every year new machines come into our orchards and groves. These machines can cut down on the costs, time, and labor per acre, making it possible for one man to handle a larger-sized operation. However, these machines like the "mechanical squirrel," hydraulic shakers, and catcher conveyer units are expensive. Purchasing equipment which will streamline your type of fruit growing operation may force you to increase the amount of land you put into production so that you can take full advantage of the labor-saving features and offset the costs.

Fruit growing requires that you locate where the climate is compatible with raising your crops. You will have to buy land near shipping centers or close in to the city. You may have to buy land near good roads and where packaging materials can be purchased easily. All these factors can affect your costs and thus the size of the operation.

A family fruit operation with small fruits like grapes, strawberries, etc. can be adapted to small farms of five acres or less. You should be able to increase production dramatically (up to about 4,000 quarts per acre) by increasing the intensity of your cultivation rather than by buying more land. One man can work about five acres except in planting and harvesting periods. In areas where irrigation is required, however, like California and the Southwest, one man may only be able to maintain about an acre.

One-family commercial fruit and nut farms usually run about the size of a quarter-quarter section, or 40 acres. Proper spacing of trees is a determining factor in producing good yields. Your state extension service will provide you with a chart showing typical planting distances and the usual number of trees planted for each acre in various orchards. Here are a few figures, taken from planting records in California, to help give you an idea of the size of land you will need.

PLANTING DISTANCES AND TREES PER ACRE *

Almonds	30 × 30	48
Apples	40 × 40	27.2
Apricots	24 × 24	76
Nectarines	22 × 22	90
Peaches	"	"
Plums	"	"
Pears	"	"
Sweet cherries	30 × 30	48
Walnuts	60 × 60	12.1

* Source: Court, Univ. of Calif. Extension.

It will pay you to check with established fruit nurseries like Starks, etc. to find out about new dwarf hybrids. These newer trees may alter your land requirements considerably. Also check with the other orchards in the area to see what fruits are most commonly grown and what trees are most commonly planted. Certain areas have a history of pest attacks, and you will want local advice on disease-resistant stock and planting distances. This way, you will learn from the other fellow's successes and failures.

If you are a beginner, don't try to buy an orchard or start one until you have taken the time to study as much as you can about the particular fruit you intend to grow. Take advantage of every expert or experienced grower you can find. The location is extremely critical in regard to topography, soils, and climate. Nothing is sadder than to see some poor dude raise a whole orchard of plums only to find that he has planted in an unsuitable location and the trees won't bear fruit! Talk about learning from experience!

Orchardmen and fruit growers will often raise chickens for eggs or broiling on the side to supplement their income while they are waiting for their trees to mature and begin bearing. Dairying and beekeeping are other common combinations. Because most of your man-hours will be invested at harvest time, even vegetable truck farming can be combined with fruit growing. Take these things into account when you determine the size of your holdings.

MAJOR PRODUCING STATES FOR
MOST IMPORTANT FRUITS AND VEGETABLES *

Citrus
Florida
California
Arizona
Texas

Peaches
California
South Carolina
Georgia
New Jersey
Pennsylvania

Apples
Washington
New York
Michigan
Virginia
California

Strawberries
California
Oregon
Michigan
Washington
Louisiana
New Jersey

Vegetables for Fresh Market
California
Florida
Arizona
Texas
New York

Vegetables for Processing
California
Wisconsin
Oregon
Washington
New Jersey

Pears
California
Washington
Oregon
Michigan
New York

Grapes
California
New York
Michigan
Pennsylvania
Washington

Cranberries
Massachusetts
New Jersey
Wisconsin
Washington

* USDA Bulletin #2221.

That Little Chicken Ranch in the West

Ever since I read *The Egg and I*, I've wanted to avoid raising chickens or eggs. Like the folks in that delightful book by Betty

McDonald, you'll have to learn as you earn with your hens. There are basically two types of poultry operations. You can allow your hens to be free part of the time, or you can keep them permanently cooped. Either way, you will not need very much land. The operation should, however, have adequate water drainage and air circulation. According to the Department of Agriculture, one hundred hens need about 8 gallons of water a day. You will probably need more to be available for sanitation purposes.

Some poultry farmers still go in for free-range brooding. This requires several small brooder houses which will accommodate from three hundred to three hundred and fifty chicks. These houses are usually equipped with a brooder stove and are moved at least once a year to promote proper soil sanitation. While it has been estimated that twenty-five birds will provide enough eggs and meat for the ordinary family of four, you will probably need to raise a flock that numbers at least two thousand hens in order to make a go with a full- or part-time commercial egg farm which will sustain your entire family. Many commercial egg farms raise from five thousand to ten thousand hens! Using good management techniques, an operation like this could be accommodated on 5 to 10 acres.

Two acres should be enough room for a small, beginner's ranch. That is, unless you plan on using part of the land for other part-time farming operations.

Locate your farm close enough to the city and good roads so that you can insure your markets. The price per acre of this kind of land is high, so seek the advice of experienced producers to make sure you have enough land.

Today, most hens are permanently housed. The Ralston-Purina Company will send you a free booklet with a feeder program for your layers, fryers, or broilers (Checkerboard Square, St. Louis, Missouri 63188). In the conventional "on-the-floor" houses, they advise that you provide 2 square feet per hen. For houses that have at least half- to full-wire floors, they recommend 1¼ square feet per hen. For caged birds, economics may determine the size of your layout and how many birds you keep, how many birds you house per cage, and what cage size you use. Producers of fryers and broilers usually plan on about 3 to 5 square feet per bird, depending on the weight of the breed.

Turkeys require a larger-sized ranch, but 10 acres will be most ample for an average turkey operation. Be careful not to go into the turkey-raising business in a big way until you know your markets. In the last two years many turkey growers overproduced themselves right out of business!

Game birds are another specialized type of poultry raising. Ralston-Purina suggests that you talk with a successful game bird producer in your state about housing requirements and the amount of space you'll need for building flight pens, a duck flight tower, or other facilities. Game bird chicks need about 1 cubic foot of air per minute for each pound of body weight.

If you plan on raising ducks or geese, the same 5- to 10-acre ranch should serve as a good rule of thumb for helping you select a farm. You can probably get by on a much smaller tract, but it's good to remember that ducks and geese usually lay better if they have a certain amount of privacy and access to a stream or pond.

Dairying

Most of the successful dairy farmers I've met say the best way to get into the business is to "grow into it." If you are a novice and don't expect to make dairying your main farm occupation, you can easily graze from ten to twelve cows on 60 acres.

Today, many dairymen have cut down on the acreage requirements by running a "dry-lot" operation. They keep their cows in loafing pens or exercise lots with sheds that are roofed, but are otherwise open to the weather. These sheds eliminate the need for the traditional dairy barns we have all known since childhood. The dry-lot operators purchase all, or most, of the roughage and grain concentrates they need for feed. This type of dairy business is typically found around our large cities. Using the cows as "milking machines," the city or suburban dairyman can operate off as little as 3 to 5 acres with a herd as large as thirty or forty.

The cows are milked in a highly sanitized milking parlor and then returned to the sheds and pens where they loaf and eat until it's time to be milked again. Running a dry-lot dairy will enable you to limit the *size* of your farm, but remember that we're talking about high-priced land within or near urban centers where taxes are high. Also, your capital outlays for modern milking equipment and bulk storage facilities will be considerable and may more than offset any savings you will realize on the purchase of land. Then, too, dairying in suburbia may not go hand in glove with your ideas of a good life in the country for you and your family.

Full-time dairying today is no different from other types of farming in that the trend is for farms away from the cities to become larger and larger. A typical dairy farm in Wisconsin is likely to be somewhere between 250 and 320 acres with seventy to eighty cows. About fifty to sixty of these cows will be producing milk at one time.

As the average farm acreage increases, you can expect to need more investment capital for land, cows, and equipment. You can estimate that your "per cow investment" on a modern dairy farm will average out at about $2,000. However, with inflation, these costs are certain to keep rising. The large dairy farm will grow most of its own feed—field corn; chopped green for silage; the first cutting of alfalfa for silage, the second cutting will be baled for hay; later, oats will be harvested for grain and bedding straw. Very little, if any, of the crops you grow will be sold. A farm this size can easily cost between $150,000 and $250,000. Be prepared to investigate various farm credit and farm loan systems available to you.

Since milk cows are generally good producers for a period of only about four years, part of the money you might want to invest in farm acreage will have to be put aside for the purchase or raising of replacement stock. Many dairymen don't bother to raise replacement heifers. Instead, they purchase freshened heifers ready to join the milk string from specialized dairy farms like the one my friend Al Bruno and his son run in upstate New York. Other dairy farmers specialize in raising replacement stock and bulls. Another form of specialization involves the raising of "dairy beef." Dairy beef farms account for about a quarter of all the beef sold in food markets across the country. Most of the dairy beef farms are feed-lot operations located near our large cities. They require only minimal acreage.

If you are not going to have to depend on your dairy operation for the major part of your family income, you had better weigh size, number of cows, and the amount of equipment you'll need *very carefully*. Like other farming activities, increased size and efficiency often turn *both* the farmer and his cows into cogs or numbers in a vicious cost-labor-price cycle. This kind of farming can become very sterile and automatic. If you want your cows to have names instead of numbers, if you want your life on the farm to have *meaning*, don't play the numbers game—and don't forget "the personal touch!"

The Size of Beef Cattle Operations

As with dairying, beef cattle operations can be tailored to many different sizes of ranches and farms. You can raise a herd that numbers in the thousands or you can buy one calf, finish it, and sell it as a prize steer. Generally, there are two main types of operations; raising beef breeder herds, or finishing feeder cattle. Beef cow-calf and breeding herds are usually raised by large ranch operators on both private and public grazing land in the fifteen western range states. Pasturage on a typical cattle ranch of this type often runs in the thousands of acres. To run a breeding herd operation, you will need to buy or rent enough land to provide good summer pasture for a large herd of steers and calves. I say large herd because the more cattle you run, the better your chances of holding costs down and getting a good return on your investment. During the winter months, you will have to provide hay and a pelletized protein supplement for the calves not sold after the fall roundup.

While some ranchers "finish" their own beef and bring it to market themselves, this phase of the business has more recently become highly specialized with the "finishing" being done on small feeder lots close in to our large cities or on typical Corn Belt farms where leguminous hay and feed grains are grown in the summer. The majority of all feed-lot activity occurs in areas where they can concentrate a great many head of cattle on very small acreage where the cattleman has to purchase all of his feed. In this instance, the feed-lot operator must be marketwise in order to buy his cattle cheap and sell them at a good price. This is a hazardous business and recent jumps in feed prices and other fixed overhead costs have "squeezed out" even some of the most experienced feed-lot operators. Consumer boycotts, protesting high beef prices, can cause serious surpluses as the feed-lot business follows the law of supply and demand.

Even though beef-finishing operations can be run on a small farm with some lower investment in labor, buildings, and equipment than other types of farm operations, be forewarned that *it is a very tough and risky business for a beginner.* One expert told me that he believes it can be documented that less than 10 percent of the beef cattle sold out of feed lots since September of 1973 were sold at a profit!

While the consumption of beef in America is not rising as anticipated, experts feel the current concern over falling beef consumption is hasty and that per capita consumption of beef and veal will hit 120 pounds or more by 1980 (USDA-ERS). They predict many small-capacity feed-lot operators will drop out of the business and the slack created by this will be taken up by the more competitive large-scale lots with a greater feed efficiency. Such large-scale lots, with head capacities of 16,000 to 32,000, have been on the increase in states like Texas, Oklahoma, Nebraska, Kansas, Arizona, and California.

Little David, Come Play on Your Harp!

Raising sheep may interest you as a potential type of farm or ranch operation. There's nothing quite so "pastoral-looking" as a country hillside partly covered with a flock of grazing sheep.

Once again, the amount of land you will need depends to a large extent on the type of sheep-raising operation you intend to undertake. Generally, you should be able to raise two to four sheep on the amount of pasture and water needed for one cow (5 fertile acres). Your sheep can be meat producers after six months.

In the leading lamb- and wool-producing states like Texas, the Dakotas, Wyoming, Montana, Idaho, etc., vegetation is thin and water is often scarce. Each sheep may require up to an acre of pasture. In these areas, ranchers may graze flocks numbering in the thousands on huge tracts of land.

Half of the sheep pasturage in these western states is on public land, rented by the sheepmen. Contrary to the picture depicted in the old "sheepmen vs. cattlemen" movies, it's a common practice for ranchers to let sheep and cattle share the same range. The sheep eat the close, soft, leguminous grasses, leaving the coarser, clump grasses for the cattle. Most of the lambs are shipped east for "finishing" when the range pastures play out in the fall.

Western sheep ranches normally require a large investment in lands. Often, this capital outlay must be supplemented with more capital to rent public lands. Large outlays for stock, equipment, buildings, and supplemental feed will also be necessary.

In the eastern states, sheep farming is usually carried out on a "small-flock" basis. Flocks are raised or "finished" for slaughter and, on a given farm, will range anywhere from twenty-five to one hundred animals.

Since a small flock of sheep can complement many types of farm operations, experts often suggest that novice shepherds begin with a small flock of about twenty-five two- to three-year-old ewes and one good yearling ram. A flock of this size can be pastured on about 10 acres.

Sheep pasturage should provide easy access to plenty of shade and water. Fields can be of timothy, bluegrass, brome, sweet clover, rape, or stubble. Rotating your pastures will prevent overgrazing and cut down on the chance of members of your flock becoming infested with worms.

In the winter, if deep snows are common to your farm area, you will need to provide housing for your sheep. Open sheds are often used for housing because the sheep do not need warmth so much as protection from drafts and heavy rains or snows. Sheds that are open at one side (away from the weather) help insure good ventilation. It's important that the flock is not crowded in its winter quarters. You should allow for about a dozen square feet per ewe and a little more than half that per lamb.

Spring is usually the most important part of the sheepman's year. This is the normal time for lambing, shearing, docking, and castrating. You should count on having to spend a great deal of your time with the flock during the two or three weeks it takes for all the lambs to be born. If you intend to diversify your farm operation into several types of activities, try to use common sense and limit the size of your flock so that the lambing will not keep you from plowing, planting, and other necessary spring farm chores.

Hogs and Pigs

Hog and pig production can be easily adapted to most types of farming operations. It's been said that even the smallest piece of country property can accommodate at least one hog. A 15-by-20-foot fenced rectangle with a simple hog house, shade, water, and a creep feeder will put you into business as a pig producer. If you buy a sow, you may need more room in short order!

Today, most producers raise pigs for slaughter. Only about 4 percent of the hogs and pigs produced in the United States are for breeding purposes. The amount of land you will need for your pigs will depend on whether you want one or two litters in a year and whether you want to raise your pigs on pasture or in confinement. Many of the most successful and most efficient producers raise their pigs in clean, modern hog houses with underground manure pits that are pumped out twice a year. Because they are concentrating their farm efforts on pigs, they opt for two litters each year. Since one out of every four baby pigs fails to survive, these producers must spend many man-hours carefully nursing their crop through the first critical weeks of life. Other farmers, who raise hogs and pigs as a sideline, allow their crop to farrow in May and June. They pay little attention to the new litter. They allow their swine to run in pasture most of the year, using portable pig houses which can be moved from field to field. It's important to move your pigs in this way in order to prevent overgrazing. Pigs that are allowed to overgraze and root are susceptible to worms, fungi, and many other soil-borne diseases.

Pigs like pastures of red clover, alfalfa, soybeans, rape, brome, etc. Studies have shown that pastured pigs need 25 percent less concentrates in their feed than confined pigs. Where good, clean pastures are available, and the hogs can be rotated to different fields to prevent overgrazing, pasturing can be both efficient and economical. You should remember, though, that pigs are strong rooters and that pastured pigs require sturdy, well-maintained fencing. Where the cost of land is high, you should think twice about running pigs in pasture. When feeding pigs in dry-lot, it's a good practice to sort them according to size and limit your feed lot to about twenty pigs. Each pig will require about 8 square feet of living space and draft-free housing where he will be warm and dry. Plenty of fresh water should also be available. Farrowing pens, 8 feet wide and 10 feet long per sow, are fairly standard.

Those Amber Waves of Grain

Anyone who has ever driven across the vast wheat fields of the Dakotas or Kansas plains knows that it takes lots of land to make our country the breadbasket of the world. The amount of acreage needed to efficiently produce the various major field crops in the United States today is astronomical!

Generally, the *more* fertile land you can purchase, the more efficient your field crop enterprise will be. In the western states of Kansas, Colorado, Oklahoma, and Texas, winter wheat is grown on huge spreads—usually the combined acreages of several 600-acre-sized farms. Spring wheat is grown on even larger parcels—sometimes thousands of acres—in the western Dakotas and Montana. It takes big money to farm the big wheat fields. Often the land is owned by major farm corporations, the leaders in the world of agribusiness. The machinery for planting and harvesting wheat is enormously costly to buy or rent. Grain storage is another cost to the growers.

With the need for so much capital outlay, sometimes "little fellers" like you and me can get a tiny piece of the action. A city friend of mine, Ken Arber, who runs a pharmacy in Omaha, owns a very small interest in wheatland acreage. From what I know of Ken's experience, speculating in wheatland and wheat growing isn't always a very profitable activity!

From the point of view of good economics, field crop farming as a full-time activity is no longer a business for small-scale operators. This is true in almost every part of the country. The small cotton farmer in the South is taking his land out of production, while the big corporation-run cotton farms in the Southwest are becoming larger and more mechanized. Washington and Oregon sugar beet cultivation is also highly mechanized as are the potato crops grown in Minnesota, North Dakota, and Idaho. The *U.S. Statistical Abstract* for 1973 tells us that the average size of cash-grain farms in the Corn Belt is continuing to get larger. Today's highly mechanized Corn Belt farm averages around 300 acres and the trend is toward even larger acreages as the small, inefficient farms are gobbled up by their larger neighbors.

Mechanization has increased farm size and farm profits, but it has also increased farmer costs. One nice thing about this is that while croplands are scarce and costly, the farmers who may want to buy their smaller neighbor's land are not always anxious or interested in purchasing the farmhouse and farm buildings that go along with it. If you are looking for a nice old farmhouse, some farm buildings, a farm garden, a woodlot, or orchard, you may luck out by taking the leavings of some farm gobbler!

Also, you may be able to do business with a farmer who would like to buy his neighbor's land but doesn't have all the necessary capital. You may be able to buy the farm yourself, lease most of the fields to the efficient, full-time farmer next door, giving him an option to buy the cropland at a later date. Part of such a deal would be for him to supply you with enough grain to feed the stock you maintain on your comfortable little place in the country.

Specialized Crops and Specialized Farming

Americans have become the most sophisticated and insatiable consumers of specialized agricultural products in the world. It seems that there is always some plant scientist tailoring old plants to new needs or bringing some relatively unknown plant into feasible commercial production. Perhaps you have a yen to design your new country-living enterprise to be able to cater to one of these new profit-making segments of the farm market. Here is a partial list of specialized crops and farming opportunities that should help you to get thinking along the right lines: African violets (require greenhouse facilities), avocados, beekeeping, blueberries, catfish, cranberries, dry beans, goats, grapes, grapefruit, mushrooms, mink ranching, peanuts, potatoes, rabbits, raisins, rice, soybeans, sugar beets, sugarcane, strawberries, tobacco, and trout.

More important than *size* when it comes to specialized farming is *location* and *adaptability* of the land in terms of soils, water, and climate for the use you want to make of it. When it comes to finding a suitable location for any type of specialized farm operation, you should take the time and pains to consult farmers already successfully in the business.

There are other ways to figure out the right-sized farm for

you. Just counting up the number of acres that you can afford to buy is not necessarily the best way to determine the right-sized farm. Farm size can be limited by the price you have to pay. Or by the amount of capital you will have to invest in livestock, machinery, building, labor, or in any combination of these resources.

For example, the place you have your eye on might have the most acreage of any spread in the world. But if it only had two cows you could hardly call it a big cattle ranch. Another cattleman might graze a hundred thousand head of cattle on public lands. His operation would have to be considered a bigger cattle ranch.

A farmer might successfully farm a large parcel of land all by himself because he had invested his money in labor-saving farm machinery. Experts would judge the size of his operation by assessing the size of his investment in machinery.

Another farmer with limited acreage might have a big investment in feed, automatic feeders, and milking parlors in order to be a successful dairyman.

Another with a large family and a number of hired farm laborers might successfully raise vegetables or flowers on small acreage using highly intensive methods of cultivation (which are often very costly).

And a man with a large family and a number of hired farm laborers might successfully raise vegetables or flowers on small acreage. His farm would have to be measured in terms of his investment in labor and by the products it produces.

Another common method of measurement is to decide on how much farm income you will need to make each year to augment your off-the-farm income and to purchase a farm that's large enough to meet those income needs. Remember though, "Don't bite off more than you can chew!"

Farm Layout May Affect Size

The way you want your farm to be laid out will affect the amount of land you buy. One of the quirks of the human mind is its insistence on giving things some formal sort of order and uniformity. This quality or, if you prefer, foible, may influence you to buy enough extra land to turn an irregular parcel into a square or rectangle.

If only part of the acreage fronts a road, you may want to purchase more frontage land in order to be able to move your livestock and farm machinery from field to field, easily. While easy road access may also increase the farm's resale value, you should keep in mind property which fronts a road will be higher priced than off-road acreage.

If a big part of the reason for moving your family back to the land is "to get out of the rat race," then there is no sense in your hopping on the new country bandwagon that's rolling faster and faster toward big, run-by-the-numbers, automated farms.

Small-scale subsistence farming may not be very practical or profitable in terms of money. But if you work off the farm too, and practice the four Ps—planning, pride, patience, and perseverance—you'll reap your rewards in peace and contentment. Your wife and kids will probably like it better in the long run if you settle on a "small pond" where you can be the "king bullfrog" and they can be the "royal family"!

The Soil Really Makes the Difference!

In your search for that "perfect" farm or country place, you will more than likely come across many places that are not so perfect. In each farm, including the one you eventually decide to buy, there will probably be certain drawbacks and deficiencies. The farmhouse may be old and lacking in modern facilities. The barns and buildings may be less adaptable to today's farming methods than they were a generation or more ago. The livestock and farm machinery may be of a quality inferior to what you had hoped to find. After a while you will come to realize that *no farm is exactly what you have in mind to buy and that you will have to make some important compromises.* You will soon accept the fact that all these problems can be licked with the help of money and hard work. However, one thing should never be compromised if you are planning to farm the land—*the quality of the soil!*

Good, fertile soil is the real difference between a profitable farm and a poor, failing farm. Rebuilding eroded or depleted and infertile soil is a thankless and expensive job. It could take you years and almost all your capital to bring worthless land back into production—so *be careful!*

My Grandma Putt always told me that the good Lord intended for us farmers and gardeners to be stewards, not miners, of the land. If she were alive today, I think she'd be pleased to see that the old, illogical practices of exploiting the land have finally been pretty much discarded and replaced with the sound techniques and methods of conservation.

But you can still find exceptions, so it's to your own best interest to see that you don't get stuck with an infertile farm. There are several inexpensive, commonsensical things you can do to prevent that from happening. The first is to get as much accurate information about the land as you possibly can.

Take an Old-fashioned Walk

I can't emphasize too strongly the fact that *you should never purchase any country property that you haven't looked at in person.* By looking at the land, I mean get out on it! Take not one but several old-fashioned walks around and across every foot of the land. If you intend to put the land to use, don't just concentrate on the acre or more where the farmhouse and farm buildings stand. That's a mistake made by too many city folks. A walk in the fields will help you get a good general idea about the fertility, drainage, methods of cultivation currently being followed, and what the best long-range use of the land might be.

From time to time, on your walks, bend down and pick up a handful of soil. Hold it in your hand and squeeze it between your fingers. Feel its texture, see if it contains moisture and organic matter. Good soil will ball up but also will crumble easily.

Many folks believe that soil must be black or dark brown before it is good for growing crops, but this is not necessarily true. The local county agent will tell you something about the kind of soils you can expect to find in the area and he'll also instruct you in the proper way to take soil samples for analysis at the state agricultural extension station. The extension service will be able to furnish you with a bulletin or circular like Circular P-9A, "Soil Testing on the Farm," which I obtained from the Alabama State Extension Service at Auburn University, Auburn, Alabama. It is excellent and gives step-by-step instructions for taking samples and keeping records. No doubt your state will furnish similar information and help.

Don't just take your walks on nice sunny days. Of course a walk in the sunshine will cheer you up, but you'll be more likely to learn more from the land in times of lousy or inclement weather. For instance, in the winter, you will see where the snow drifts. After a heavy rain in spring or fall, you will see how the water runs off or where it stands in puddles, indicating poor surface drainage. After a particularly heavy rain, you may discover that part of the land is situated on a flood plain! On a windy day, you might find out that there are drafty corridors on the property that might harm crops or carry fire from a nearby forest or woods. Prevailing winds can also cause severe erosion. On a foggy day, you might find damp pockets on the property, places where your crops or orchards might be destroyed by frost.

Know the Soil *Before* You Buy

Information about the soil on a farm site should not be very difficult or expensive to come by. I'd suggest you take the time and make the effort to know all you can about a farm's soil and topography *before* you buy it. Most farmers are proud of the steps they've taken to insure the long-range productivity of their land.

You, as a prospective buyer, shouldn't hesitate to ask the seller some pointed questions about the conservation methods he has employed during his tenure on the farm.

Ask him if he has had the topography of his fields surveyed and the soil tested. If so, find out how long ago that was and ask to see the reports. Find out how well he has followed the correct land-use procedures suggested by the soil specialists who conducted the survey.

Ask him to show you his farm records for past years so you can see what crops were planted where, and in what kind of rotation. Find out what amounts of seed and fertilizer he used; what kind of plowing and planting system he follows in order to retard erosion; and what yields-per-acre he has been getting.

Find out if the farm has ever been troubled by infestations of nematodes or any other soil-borne or soil-debilitating diseases. If you find that the seller is the least bit reluctant to discuss these topics thoroughly, *you should be wary!*

Many farmers have no idea whether or not their land has ever been surveyed. They might be surprised to learn that it probably has. Back in 1935, the federal government set up the Soil Conservation Service (SCS) during the worst drought in our nation's history. Ever since then, it's been the job of SCS scientists to aid farmers and other citizens who use the land to protect it; use it fully and properly; and to preserve its utmost productivity for future years. By now, almost all the productive farmland in the United States has been analyzed as part of the National Cooperative Soil Survey.

To find out if the land you are interested in buying has been surveyed, contact the Soil Conservation Service, U.S. Dept. of Agriculture, Washington, D.C. 20250. Often, you will be able to find a copy of the survey report at the local library. If they do not have a copy of the survey maps and report, check the county agent or see if there's a district office of the SCS nearby.

If you wish to obtain a copy of the survey report for your own use, you can do so for a minimal charge by requesting it from the

Superintendent of Documents, Washington, D.C. 20402.

The Soil Survey Report combines the results of aerial photomapping, on-location field work, and laboratory analysis of soil samples. All the data that has been gathered is evaluated to determine soil types, slope, extent of erosion, availability of water, drainage, and other characteristics.

To take the representative soil sample on the farm, the soil specialist probably waited until conditions were just right for plowing. About twenty borings are taken in each field. (A field is usually defined as a fenced-in area confined to the raising of a single crop.) A sharpshooter spade, soil tube, soil auger, or long-bladed trowel can be used to take the samples.

The important thing about taking samples is to be sure to take a uniform slice of soil with each boring. This slice should include portions from the topmost inches of the surface soil down to a depth of about three feet. The twenty borings are then mixed together in a clean bucket to make a composite sample. A jarful of the composite will be enough for the lab scientists to determine the character and class of the soil in that field.

As I've said before, no two farms are alike. After looking at a soil survey report, you may rightly conclude that no two *fields* on the same farm are precisely alike!

The National Cooperative Soil Survey has classified more than eighty thousand different types of soils! Character and class of soils are generally graded in terms of: *clay loam, silt loam, loam, fine sandy loam,* and *sandy loam.* Loam, which is about halfway between the heavy, slippery clay and the coarse, gritty sand, is the best texture of soil for growing plants. However, many crops will grow in silt loam or even sandy loam if the proper cultural practices are followed.

Nutrient Content

Although it is very difficult to determine the amount of available nutrients in soil, a number of chemical and biological tests will be made. These should indicate if there are adequate amounts of nitrogen, potassium, and phosphorus. Tests for trace elements like boron, calcium, chlorine, manganese, iron, etc. are usually not made unless an obvious problem exists or a specialized crop is being grown.

The test taken should indicate whether the present owner has used adequate liming and fertilizing methods. The report, or guide, will indicate fertilizing methods and quantities that should be followed in order to get good harvests.

Acidity and Alkalinity

In order to release the valuable mineral nutrients and make them available to growing plants, the soil must have a proper balance between acidity and alkalinity. Most soils in dry or desert areas are alkaline. Most soils in moist and humid areas are acid. It's fairly easy to find out the amount of acidity or alkalinity in a given soil sample by giving it a pH test, potential of hydrogen. It is measured on a scale of 0 to 14. The lower the pH, the more acidity in the soil. The higher the pH, the more alkaline the soil is. The closer a given soil sample tests to neutral pH—or pH 7—the better the acidity/alkalinity balance.

A few plants flourish in soil that is "sour" or acid. Azaleas, blackberries, potatoes, sweet potatoes, oats, and rhododendrons are good examples. Other plants are sensitive to acidity and only grow well in "sweet" soils that have a high pH, or that have been heavily limed to increase alkalinity. Among these are: alfalfa, celery, onions, spinach, sugar beets, and roses.

It's important to know the pH preferences of the kinds of crops you intend to raise. If they prefer soil that is alkaline, and the soil at a given farm site is very acid, it may not be worth your while to try to tip the balance dramatically with heavy liming. Instead, you should change your plans and grow acid-loving plants or plants that could flourish in the soil if *average* applications of lime were made. If you have made up your mind to grow only certain crops, and the acidity/alkalinity balance is improper for them, then *pass up this farm and find another*.

In order to simplify the soil survey reports for people like you who will be using the land, the results are summarized in an "interpretation." This summary puts the soils and the land at a given farm site into broad groupings or classifications which indicate how well they can be expected to grow specific field crops or pasture; whether they are suitable for range; for timber; for drainage; or for irrigation. You will probably be most interested in the interpretation of the farm's *land-capability classification, productivity estimates,* and *management suggestions.*

In the report, the individual soils of the surveyed area will be matched against eight standard land-capability classes. These are graduated on a decreasing feasibility scale—from Class I, which is capable of growing all the cultivated commercial crops common to the climate, to Class VIII, which will not support any commercial crop.

We should all be so lucky as to find a farm with Class I land! It can be used to raise crops, for pasture, range, wildlife, or woodland. Class I land doesn't require that the man who owns it introduce any expensive conservation measures in order to use the land fully and get the best return profit for his investment.

Although the lands that fall into Classes II to IV are still suitable for profitable commercial farming, each class requires increasingly more costly management and conservation practices by the farmer in order to protect the quality of the soil and keep it productive.

Classes V to VII are considered too expensive to farm, but are often used for profitable livestock grazing or forestry; Class VIII land is not much good for any kind of farming, but perhaps it can be used as a refuge for wildlife. On a given farm site, there may be land that falls into several of these land-capability classes. If there is a small parcel of Class VIII land on the property, you may want to consider using it as a wildlife conservation area. The Fish & Wildlife Service of the U.S. Department of Interior, Washington, D.C. 20242, will furnish information on how this may be accomplished.

Remember the scene in the movie *Easy Rider* where the kids living in the commune tried planting on a parched piece of southwestern desert? Well, that was Class VIII land. What those young people needed, even more than faith and hope, was a soil survey! Before you settle on that "piece of promised land" you've picked out, find out if the earth is capable of rewarding your labors.

If there is a local SCS office in the area, it's a good idea for you to have one of the soil specialists there to interpret the survey report and evaluation guide and maps for you. If not, ask your county agent to help. Ask the specialist to evaluate the methods and practices that the current owner is following and the correctness of the ways he is putting the land to use. If his methods are bringing disaster, *now* is the time to find out, instead of later, when you will have to spend a fortune to correct his mistakes.

Tell the soil specialist your plans for the land. Find out if your plans are economically feasible. Have him describe the various alternatives so you and he can develop a long-range conservation plan that will insure that the land remains profitably productive for as long as you and your heirs hold it and far into the future.

This conservation plan will be designed to include every acre of the farm, ranch, or woodland. It will help you, *the user*, to get off to the right start and catch you in time if you had been getting set to begin what, in the long run, would turn out to be a losing battle.

It will help you, *the farmer*, employ proper cropland conservation practices to control erosion, save water, and establish permanent vegetative cover. These measures may include: contour farming, strip-cropping, stubble or trash mulching, terracing, shrub and tree planting, and pond constructions.

It will help you, *the rancher*, set up the correct grassland practices to improve the quality and condition of your range and to balance the size of your livestock operation with the availability of your forage and water resources. There is no point in your having more cows or sheep than the land will support. It will show you how to do this by implementing such techniques as: rotation grazing, pasture planting, deferred grazing, and range seeding.

If the land you are looking at is woodland, the guide will help you, *the forester*, evaluate the quality and condition of the stand. The plan formulated by the SCS scientists will let you harvest wood crops successfully by the proper methods of: improvement cutting, tree planting, windbreak planting, etc.

If you intend to concentrate on recreation farming, or just allow the land to lie fallow, the specialist will design a plan to hold the soil and prevent erosion while building up fertility. It will help you, *the recreational farmer* or *conservationist*, to create and promote the proper habitat for wildlife. Recommended practices may include: cover cropping, green manuring, hedgerow planting, fish pond or stream improvement, and wildlife area improvement.

No matter what your farming intentions, the evaluation guide may show that the land has poor drainage and needs to have tiles installed. Or, that it is in need of a reservoir and properly laid-out waterways for irrigation. Obviously, these are all things you should know about before the farm is yours.

These are all things that can be followed successfully or corrected if they are not being followed already by the present owner. Certainly, what is being done with the land now and what should be done with it are important pieces of information which should affect the market value of the farm as it is. This information should certainly affect your decision on whether or not to buy.

Watch out for Abandoned Farms!

Be extra wary about buying a farm that has been abandoned or has been out of production for several years. Even though such a farm may appear to be a great bargain, you should not be anxious to snap it right up. Chances are that if it has been abandoned for a number of years, there is a good reason. Maybe the land is too eroded or infertile to make farming it a sound commercial venture. Maybe the reason that the previous owner gave up was because the soil became infested with nematodes. These are microscopic microorganisms which will destroy almost any type of crop you might want to plant. Maybe the water source is inferior or the drainage poor because of a layer of hardpan just below the topsoil. Maybe the farmhouse is rotten—eaten away by termites. Maybe the house and farm are constantly being flooded out. Maybe the federal government is leasing adjoining lands to a logging or strip-mining company and they will destroy your watershed and your peace of mind. There could be a thousand good reasons to pass this "big bargain" right by!

Section Three
The Farmhouse, Utilities, and Facilities

Up to now, I've asked you as a prospective farm buyer to look at where you are going to farm, what you are going to farm, how big your farm operation will have to be, and what kinds of soils you will be farming. But you're probably wondering why I haven't paid any attention to the one thing that's most important of all to you—"Where is my family going to live!"

Well, sure, the farmhouse is important. No doubt about it. But the reason I've ignored it up to now is that it's almost always the first thing that city people who want to move to the country look at. Sometimes it's the only thing they consider when they make the all-important decision of whether or not to buy a piece of country property. I hope that by delaying this section just a wee bit, I've been able to show you that buying a place in the country is not like buying a house in the city. This is especially true if you intend to do something with the land. As tough as this is to get across, *it is far better to purchase good land with a poor to mediocre farmhouse—or even no farmhouse at all—than it is to purchase a great farmhouse that's perched on poor to mediocre land.* Sometimes the sight of a beautiful farmhouse, set up on a tree-shaded hill, makes folks overlook faults in the rest of the property that would turn off even the most optimistic farmer with a little bit of experience. Many people come to the country in search of a home that offers more in the way of comfortable living conditions than the quarter they've been occupying in the city. They might be unpleasantly surprised to learn that according to the U.S. Census, nearly two-thirds of all the substandard housing in this country is rural housing.

Those beautiful old farmhouses you pass along the highway may look bigger and more livable than what you're used to, but chances are they are all broken up inside into tiny rooms and that many of their facilities and features are obsolete. When most of them were built, it was customary for farm families to be very large. Several sons were needed to work the land. It wasn't unusual for one or more brothers to grow up, marry, begin to raise a family, and continue to live on the same farm with their parents. Even when something like that didn't happen, it wasn't unusual for a farm family to have city cousins or nephews, like me, come to live with them and add to the cheap labor supply.

I can remember sharing one of those tiny upstairs bedrooms. They were so small, even an eight-year-old had trouble turning around without bumping into something or someone. Funny, with all those supporting and nonsupporting walls you'd think that the plaster would never crack or that the floors wouldn't creak or be drafty. But I swear, that wasn't true in either case!

When you look at an old farmhouse, you'll want to carefully check the construction to see if the floor plan is adequate, or can be easily altered, to fit the living and sleeping needs of a modern family such as yours.

Somehow, that same large farm family I remember so well functioned without the benefit of hot and cold running water in the house. A hand pump was attached to the kitchen sink. It was connected with the well outside. Seemed like we pumped for an hour to fill a dishpan or bucket. It took several buckets of boiling water to fill the bottom few inches of the Monkey Wards bathtub on Saturday nights. Seemed like it was always cold before I got to use it. That's right, we had to share the bath water! That is, unless we wanted to pump another hour's worth of water and wait for it to boil! In those days, bathing was a once-a-week *chore*—and you didn't get to luxuriate in the tub.

While I'm on the subject of the plumbing, I should mention that those trips to the outhouse were mighty chilly on a cold winter's morning! That's a fact which caused several generations of rural Americans to have problems with irregularity! The way I remember it, *bathing* wasn't the only once-a-week chore!

If time, distance, and nostalgia have softened the memory of what it was like to boil your water on a black iron stove . . . to hop out of bed onto icy linoleum and race to the floor register so you could thaw your feet . . . to have your cold fingers fumble with matches at the kerosene lamp . . . so you could see to go and find the chamberpot—which was always under the wrong side of the bed—even if you can remember all those things with a smile, I'll bet you'll want to check out the *heating, electricity,* and *plumbing* carefully so you won't ever have to live through those "good old days" again!

How to Inspect an Old Farmhouse

Prospective country home buyers often ask, "What should we look for when we inspect a farmhouse that we're interested in buying?" A proper answer to that question would produce a list as long as your arm! Perhaps the best advice I can give you has to do with *your attitude*—on *how* you should look, rather than on *what* you should look for first.

In most cases, the existing farmhouse was built on the best site available to locate the home. If the place is decent-looking and repairable, there's no reason why you shouldn't turn it into the home you've always dreamed about. You and your wife probably already like something, or some things, about the appearance and location of the old place or you wouldn't be taking the time to inspect it more closely. But remember, an old house *is* an old house. In most cases, the electricity, heating, and plumbing in an older house leave a lot to be desired. All three were usually installed many years after the house was built. The farmer living there at the time may have done some or all of the work himself. Being a good farmer does not automatically make a guy a good electrician, plumber, or carpenter!

The first thing you both should do is put on your "we're-from-Missouri-so-you'll-have-to-show-us" faces and form a *negative* fact-finding team. Very few of us like the idea of approaching a new experience from a negative point of view, but buying this country home may turn out to be one of the biggest decisions your family will ever make. I suggest a *negative* approach because what you should be looking for, basically, are those "fatal flaws" of the house which will be so costly to repair or correct that the place will have to be crossed off your list as a possible buy.

The first time you tour the house, ask the seller and realtor to join you. They may be able to point out many things that escape your notice. Try to be objective and impassive as you evaluate each feature in their presence. Both the seller and the realtor will be searching your faces and listening to your comments to see if they've got you hooked. Too much expression, either positive or negative, is likely to make the seller react in a way you don't want. Be polite if he proudly shows you some cabinets or shelves he built himself (even if they aren't so super). Compliment his wife on the wallpaper she selected and hung herself (even if it looks dreadful). Above all, don't point out every flaw or you'll turn him against you and he'll be tough to deal with when you negotiate. On the other hand, if the seller likes you, he may throw in some valuable furniture, appliances, or tools that he had not planned on including in the deal. If you do end up buying the place, you may want to call him up some day and ask a vital question. It just never hurts to make friends. So be friendly but impassive about the house. Quietly make note of all good and bad points about it. At the end of the tour, if your overall impression is a good one, thank everyone for being so helpful and ask if you can come back at another time.

The next time you look at the house, bring along a tape measure, pocketknife, pad, and pencil. The first two tools will come in handy, and the taking of a few notes along the way may prove to be invaluable when you begin to bargain to get the price down. This time, try to take your inspection tour without the owner coming along. No point in getting him nervous as you poke into every nook and cranny. I suggest you begin this inspection tour of the premises on the *outside* of the house.

How Firm a Foundation?

Take a slow walk around the house. Observe the foundation. Be on the lookout for any large cracks that may indicate soil settling which can cause the roof to sag or the floors to slope unevenly. You will probably see plenty of hairline cracks in the foundation, but you can ignore these and concentrate on finding big cracks which your finger or the blade of your pocketknife will enter easily. In addition to indicating faulty construction, these cracks can be a tipoff to a damp and leaky basement or some other moisture problem inside.

It's the job of the foundation to support the weight of the walls and roof—even when it's covered with a heavy blanket of snow. The foundation must also withstand the force of pressure caused by groundwater and by heavy winds. It must also retain heat inside the house. Foundations that are bulging, or which have serious cracks, are often sitting on inferior footings placed on soil that is settling. Check the soil around the foundation. If it is extremely clay-ey, it could cause problems. Clay soils swell when wet. They can exert awesome pressure against the foundation and basement walls during periods of heavy rainfall. Soils that are light and sandy may not be able to support the weight of the house properly. These soils may also be the cause of slopes and sags. If the house has been built on boggy land, it could disappear altogether!

If the house is not firmly or properly anchored to the foundation, strong winds may have caused it to shift. This will have created a strain on the unsupported walls which in turn probably have produced cracks in the plaster in the interior. This is a common problem in country homes that were built where there were no local or county building codes.

Water pressure can also displace a poorly anchored house. The result will be numerous and costly problems that make it good common sense to cross this house off your list.

While you are outside, check the slope of the land away from the house. If you are lucky, you will find a house that sits on land that is level and has a slight downward slope as you walk away from it in any direction. This slope will insure proper drainage of groundwater after a heavy rainfall.

Be wary of that house perched up on a hill like a diamond in a ring setting. If the slope is too steep, you may have trouble with the ground heaving in the winter, causing slippage or cracks in the foundation. You may find it difficult to grow a nice lawn, to keep the topsoil from running off during the rainy season. Worse yet, you may have trouble finding the right kind of soil percolation for your septic system to operate effectively, or the proper water draw for your well pump to work right. Consult the county agent, county engineer, or county health service on how to get a percolation test made of the soil.

Sometimes a hilltop house will look great in the middle of the summer, but you may find that groundwater tends to "pond" across the lower driveway in the spring and snow drifts up at the same spot in the winter. It will get icy in that location too, making it almost impossible to get a car up to and down from the house during those periods of bad weather. My friend Tom Azzari and I saw a beautiful house like that near Aurora, Ohio. Tom said he'd sure like to own a home like that. The next time we saw it was after a heavy spring thunderstorm. The owner was crossing the pond over his driveway in a rubber raft!

While walking around the outside of the house, look up at the roof overhang, the gutters, and the downspouts to make sure they are in good condition. If the wood in the overhang is rotting, make a note, as that is expensive to repair. Look to see that the downspouts are arranged in a way that they will efficiently carry rainwater and melting snow off the roof and away from the foundation of the house. If the downspouts deposit water too close to the foundation, you will have to lay down some concrete splash blocks to correct the problem. A new guttering job could cost quite a bit of money. If the ones you see are rusted through, in disrepair, or in need of paint, be sure to make a note of it.

If a ladder is handy, climb up and look at the roof. If you find that large portions of it show wear, decay, or water damage, you can almost be certain the roof leaks. The cost of roofing materials has more than doubled in the past two or three years of this current inflation, so if the roof leaks, try to get the seller to incur the cost of having it fixed.

Check the metal flashings around the chimney and the chimney itself for any places that aren't sealed properly and which will allow water to leak in or cause costly heat loss during the winter. Note any such problems. If they are serious, ask the seller to have them repaired.

The greatest amount of heat loss from a home occurs through the roof. For years, roofing salesmen have used this fact to convince homeowners to reroof their homes and to insulate their attics. Bob Stivers, the famous TV producer, tells about when he sold roofing and insulation in his youth. He says he was always delighted whenever there was a heavy snowstorm because it was a sure sign that business was going to pick up! On the first sunny day after the snows, Bob would simply drive up and down the streets of a neighborhood until he spotted a house without any snow on the roof. He'd call the homeowner outside to see for himself (usually herself). What the homeowner saw was alarming! Almost all the other houses on the block had a thick blanket of snow on their roofs. In comparison, her roof was bare—the snow had melted. She could see the steam rising like smoke off the wet shingles! Then Bob would go into his pitch, telling her that her home was losing costly heat from inside the house (at this point, he'd dramatically gesture toward the steaming roof). He would then explain how new roofing or better insulation in the attic ceiling would prevent such heat loss and cut her heating bills by 15 to 25 percent! He says he almost always had a sale.

If it's wintertime and has snowed recently, you should look at the roof to see if the old farmhouse is properly insulated. If not, maybe you can use Bob's old sales gimmick to convince the seller that the job needs to be done and that he should knock the cost of installing insulation off his asking price. Good luck!

Now before you go inside the house, check the exterior paint job for scaling and peeling. No point in your having to pay for painting the place too. Peeling paint can be a sign of wet wood. Examine peeling spots for dampness in the walls or siding. Is this caused by rotting?

Also, try the windows to see that they're not warped. If they are locked, test them for security and make a note to try the windows on the inside to see if they can be raised and lowered easily. Is the putty holding the glass in good condition? If the glass is loose, you have found a place where the owner is losing valuable heat. If this is an area where the winters are severe and the summers are hot and humid, find out if storm windows and screens exist for the house and if they are in a good state of repair.

Now brace yourself and go inside. It makes very little difference whether you start in the attic or in the basement—just as long as you hit both places in your tour of inspection. As you look the house over from top to bottom, or from bottom to top, measure the size of each room with your tape measure and draw its floor plan in your notebook. Mark in closets, cupboards, major fixtures, and appliances. When the tour is complete, you should draw a complete floor plan of the house. You will refer to this many times before making your final decision. I'd suggest going down to the basement first because if you discover any serious problems there, it may become unnecessary for you to climb up into the attic.

The Basement

As you go down the basement stairs, check the headroom. Do you have to bend down to avoid bumping your head? Over a period of years, this could get to be a pain in the head—or the you-know-what!

Is the cellar stairway shaky or in need of repair? A farmhouse basement or cellar often looks like the Black Hole of Calcutta, so don't expect too much. The lighting may be poor, so take a flashlight along.

Remember those cracks in the foundation you found outside? Consult your notebook for their location and try to find them on the basement wall. They may have been patched and covered with a recent coat of waterproof paint. This is not always an effective way to repair leaks. Look closely to see if the painted surface shows any signs of seepage at the spot where you found the crack outside. Patching from the inside is rarely effective, even with the new epoxy-based paints that are supposed to be watertight. On the other hand, digging down in the soil around the foundation and applying waterproof plaster can become an expensive proposition. My friend Willy Williams says you can take it from a realtor who has seen thousands of homes, "Leaky basements always end up leaking again no matter what you do to repair them."

So inspect the basement walls closely for waterlines. These are strong evidence of previous flooding. Look for a sump pump or drain system that's been recently installed. Either one of these is a sure sign that water in the basement has been a serious problem. Ask the seller and his agent about it. Remember, the Law of Agency requires them to make full disclosure of any serious problems that they know anything about. If they minimize the dangers of seepage, moisture, or flooding—or deny them—ask if they'd mind giving you a statement to that effect in writing. With that kind of written guarantee, you might have some recourse in the courts if it turns out you were misled.

While you are down there, look at the floor. Is it concrete or dirt? Cellars with dirt floors are almost always slightly damp. They are sure to leak in times of heavy rainfall. If the floor is cement, look for cracks which indicate settling.

See if the basement windows are in good working order and above grade. If they are below grade, or ground level, they will provide a perfect channel for water to enter in the rainy season. Or, in winter, when snow drifts up against them, it will melt and allow dampness to seep in. It's extremely important that these windows are in good repair and watertight.

If there are one or more outside cellar doors, try them. Do they open, close, and lock easily? Are they watertight? Is one of them large enough to let you carry appliances, tools, and storage boxes in and out easily?

Now, check the size and overall layout of the cellar. This will help you decide whether it will be able to serve you as a tool room, a place for your family to work on their hobbies, a laundry room (if none is available upstairs), or as a safe, dry place for storage.

Many old cellars are dominated by the central presence of a great furnace. It's probably a coal furnace with big, elephant-leg-sized forced air pipes reaching out in all directions like a giant octopus. My Uncle Bob has such a furnace, but it no longer uses coal as the fuel to heat the house. It has been converted and a compact oil burner is now housed inside of it. In some areas, these old furnaces have been converted to natural gas. (More about checking out your heating system later in this section.) Right now, just examine the furnace in terms of the amount of space it occupies. Can you move around down there without getting a skull fracture? Does the size of the heating setup make it impossible to use the space for any other purpose? Has conversion to oil or gas freed a coalbin area which can now be cleaned, painted, and turned into a cold cellar for canned goods or a work room?

Next, inspect any posts, columns, or pillars used to support the floor above. If they have settled or deteriorated, will it be difficult to fix them?

Turn your flashlight up and check the ceiling. Is there plenty of headroom? If there is no ceiling, look at the studding under the first floor. Are there signs of unevenness? Does the floor above slope or sag? Is it warped? Does it need reinforcement with posts or shoring? Can these problems be easily and inexpensively corrected?

Run your hand over the exposed surface of the wood. If it feels slightly moist to your touch, jab at the darkest corners with your pocketknife. Conditions are probably ideal for a group of parasitic fungi which feed on damp wood. This group also included the misnamed "dry rot" fungi which carries its own water to the job and works on dry wood. Also, check this wood for termites, carpenter ants, post beetles, and woodborers. If you find *any* suspicious-looking insects (including bees) working on the wood *anywhere* inside or outside of the house, try to capture one or two specimens for identification by a pest control expert, your county agent, or a forest service officer. Those tiny little critters have turned many fine mansions into sawdust. So find out if there is a problem *before it's your problem!*

Many states and counties require that the seller furnish the buyer with a copy of a "Termite Inspection Report," which shows that the house is free of such pests. According to law, he must do this before a sale can be completed. Even if there is no such legal requirement in this locality, don't hesitate to ask for such a report as a condition of the sale.

If an expert tells you there is a serious termite, insect, or fungi problem present, scratch this house off the list. Or ask the seller to repair all damaged areas and hire an exterminator who will guarantee to rid the house of the pests, *in writing.*

Along similar lines, look for tracks or nests of any other pests that could turn out to be a nuisance after you buy. Silverfish are often attracted to damp cellars. They can easily be removed by spraying.

Rats and mice are another problem. If you see any signs of their presence on the premises, have the seller hire an exterminator who will guarantee in writing to make the house rodent-proof.

If it turns out that the basement is dry, clean, and has been converted to a modern playroom/laundry room combination, you are all the better off! Some country houses don't have a basement. Or they may have only a partial basement with a crawl space under the remainder of the house. In either case, wear some grungy old clothes, take your flashlight and knife, and crawl in there to poke around as described above. Don't get an ulcer later because you didn't want to be uncomfortable when you inspected the place. Consider all these things as pluses or minuses in the overall rating you give the house. Now, you'd better get upstairs before the owner begins to wonder what you and your wife are doing down there so long!

What to Look for in the Attic

If the basement or cellar doesn't have too many flaws that will be too costly to correct, you can proceed to the attic, as it is the next likely place to be in need of major repairs. Most country attics have no electricity, so be sure to take your flashlight along.

Once you get up there, be careful not to step between the joists or you'll put your foot through one of the seller's bedroom ceilings on the floor below. That could prove slightly embarrassing! If the attic is partially or completely floored, this won't be a problem.

Look up at the roof from the inside now and try to locate any holes or crevices where the rain, snow, or moisture gets in. See if the inside of the roof has been insulated and, if so, if there are any moisture stains on the batting. One of the best types of insulation has an aluminum foil backing which will create a vapor barrier to hold heat inside the house and keep moisture outside.

If the roof appears to sag, make a note. This could be very difficult and costly to repair. Has the floor of the attic been insulated? This could have been done with either a loose rock wool or with the roll-type insulation with the foil backing. The latter will prevent moisture from reaching the plastered ceilings below and provide a barrier to keep heat downstairs where it will do your family some good. Is the attic heated? If so, it may be possible to convert it into an extra room for writing, studying, or sewing. It's always nice to have a room like this, away from the rest of the house, where family members can go to get away from it all and be undisturbed. If there is no electric wiring, how big a job would it be to run some lines up?

Good ventilation will insure that the attic remains dry. Look for louvered vents at each end which are slanted down to keep the rain out, but which otherwise are always open. Some attics have suction fans which help to cool off the whole house during the summer. If this one does, you have to consider it a big plus.

In California and some other places in the South and West,

tree rats commonly frequent attics. In other places, squirrels and woodpeckers like to make their homes there. If you find any signs, it's exterminator time again!

Use your pocketknife to check exposed wood for dry rot or termites. Try to judge whether or not the lathing for the ceilings below are in good condition and level.

The chimney can provide an easily accessible channel for water or moisture to reach almost any part of the house, so check the chimney carefully. Usually, if moisture is present, it can easily be eliminated by replacing the metal flashings out on the roof, then tarring them over until they are secure and watertight. Sometimes if the rain has been running down the chimney for years, it can damage the masonry work to the point where you, or the seller, will have a major rebricking or patching job on your hands.

The Second Floor

On the second floor (most old farmhouses have two stories), you will first want to look at the overall layout to see if it is suitable for your family. Are the bedrooms designed to allow some flexibility in your furniture arrangements? Which is the master bedroom? Do you like it? Are there enough closets? Old farmhouses are often lacking in closets and storage space for towels, linens, and the like. There should be adequate storage areas convenient to baths and bedrooms.

Look at the upstairs bath or lavatory, if there is such an animal. Examine the tub, sink, and toilet. Are all the parts in good working order? Is the water pressure adequate? You won't want a bathtub that takes an evening to fill. Location of the bathroom is also important. Hopefully, it will be off the hallway at the top of the stairs where it's easily accessible to family members and house guests. Nothing is more inconvenient and sometimes embarrassing than to have to track through someone's bedroom in order to get to the john or to take a bath!

Is the upstairs all broken up into tiny rooms? Perhaps you will want to remodel it so that it's more open or more fitted to your needs? If so, you will be smart to come back later with a competent builder who specializes in such work. He will have a good idea of what it will cost you to do what you want. He will also know if there are any local or county building codes and restrictions which will make remodeling the house a major hassle.

As you leave the second floor, check the stairway as you are descending. In many old houses the stairways are very narrow. Measure the width to make certain you can get beds, dressers, and other bulky furniture up and down them without too much trouble. Also, check to see if any treads are worn and need replacing.

The First Floor

On the first floor, start at the front door or main entrance and step outside. Go down the walk, then turn around and come back. Is the front entrance to the home attractive and serviceable? (Inside the front entrance, it's convenient if there is a closet.) See if the door fits properly and is secure when locked. Does the door bring people directly into the living room. Is there any entryway inside or a covered porch outside? Back in the "good old days," family members used to sit out on their front porches on warm evenings and *actually talk to each other!* Today, that idea is a bit of America's vanishing folkways. I guess it was television that took our families off their front porches. Pity.

How does the house feel to you when you walk in? Is it warm and inviting? Can you smell something nice cooking in the kitchen? My friend Sophie Kantor tells me that whenever she wants to sell her home, she tries to have a spice cake baking when potential buyers go through. Sophie says the cake makes the house smell like a nice place to live! You might want to try Sophie's little ploy on prospective buyers of that house you are trying to sell back in the city.

The Living Room

A couple of generations ago, farm people rarely used their living rooms. They kept them closed off to save on heat and only opened them up and aired them out when the preacher came to call or someone in the family got married or died.

Nowadays, most of us make the name, *living room,* fit the circumstances. Get out your tape measure and put the dimensions on your floor plan. If you have a large family, make certain the living room is spacious enough so you can all fit comfortably with some friends and still keep furniture and people away from normal traffic lanes.

Check plastered walls and ceilings for cracks. If you see any that are severe, they will have to be dug out and widened, then filled in with patching plaster and repainted. Cracks in walls and ceilings made of wallboard, or drywall, can be filled, sanded, and taped before repainting. Or, if they are too wide to fix that way, you will have to replace the drywall. Minor cracks that would show after painting over them can be covered with an attractive wallpaper.

You may want a living room that has a fireplace. If this house does have one in the living room, see if it's drafty. Mentally figure out where you would store firewood and how you will bring it into the house without tracking all over the rug. Nobody wants to have to walk through snowdrifts in order to be able to toss another log on the fire.

Check the living room for electrical sockets. You may want to use the room for TV viewing, hi-fi listening, or for reading the newspaper in your favorite easy chair. It's tough to do any of these things if there aren't enough wall sockets to go around.

Inspect drapes and carpets to see that they are clean and not worn out. If you are a hardwood floor freak, it's difficult to know what you are buying if the living room has wall-to-wall carpeting. Ask the seller about the floor. When he is out of the room, you might try a little "jump test" to see if the floor is solid and doesn't have warped or saggy spots where it groans or creaks.

If this farmhouse has one or more downstairs bedrooms which you won't be needing for that purpose, see if it could easily be converted into a den, family TV room, or farm office. That's a plus.

A Downstairs Bath

If there is a downstairs bathroom, look to see if it has been located with a mind for privacy or convenience. You won't want everyone in the living room interrupted whenever someone uses the bathroom. You will, however, want a lavatory and bath or shower that is easily reached from every room. If this is the only bathroom in the house, it shouldn't be located near the front of the house.

When inspecting bathrooms, check for loose tile and a poor caulking job around the tub or shower. Flush all toilets. See how long it takes to refill the flush tanks. See if the faucets in the sink close tightly and do not drip. Examine the floor tile to see if any are loose or badly worn.

A Separate Dining Room?

If you like to have the entire family sit down together for at least one nice meal a day or if all your friends and relatives from the city will be coming out on Sundays and holidays for one of your country dinners, you may want a house with a separate dining room. Ilene calls them "formal dining rooms," and I guess she knows what she is talking about—although we have one and I've never worn a tux to dinner!

If the farmhouse has one, measure it. See if it will accommodate your dining room furniture plus all those hungry city folks and your whole family? Even with that kind of a crowd, there should still be about 3 to 4 feet behind each chair so folks can relax and for someone to serve conveniently. And be sure your dining room is adjacent to the kitchen!

The Farm Kitchen

No room will be more important to your family farm operation than the kitchen. This is the nerve center and natural gathering place for all the members of the family. This is where you and your children will start and end each day's activities. Therefore, it is extremely important that of all the rooms in the house, you inspect it carefully to make certain that it is spacious, cheerful, convenient, and comfortable.

The eating and seating area for the kitchen table should be roomy enough to allow easy access for family members and for those who serve them. Figure out the amount of square feet you will need for each family member plus one or two extra. Then measure to see if the present kitchen layout provides enough room. It's an added plus if there is a window nearby where you can all sit and look out at some of the natural beauty that makes living in the country so worthwhile.

My favorite home design expert, Sonya Selby-Wright of "The Mike Douglas Show," has just remodeled the kitchens in her home in Philadelphia and her farm in Maine. She tells me that she favors a U-shaped work area with the double sink at the bottom of the U and the stove and refrigerator across from, but not facing, each other on the opposite legs of the U. Sonya says this setup is most logical because it lets the housewife work out of the way of normal floor traffic. The work area that is all lined up against one wall is the least utilitarian because the person doing the food preparation and cooking has to take many unnecessary steps back and forth as she works.

More convenient are the L-shaped and the parallel setups. If the kitchen you are inspecting has a parallel setup, use your tape measure to determine the distance between appliances on one wall and the counters on the other. Sonya recommends 4½ to 5 feet minimum between facing equipment.

Many farm kitchens do not have enough wall cabinets or pantry areas. These are essential. Often, cabinets are hung too high and it is almost impossible for an average-sized woman to reach the topmost shelves without a stool or small stepladder. Cabinets hung about 15 inches above the normal 36-inch high counters will eliminate the need for ladders or stretching.

If any of the kitchen appliances are offered as part of the sale, check them out carefully to see if they are in good working order. If not, and they are not easily repairable, ask the seller to remove them when he moves. You don't want to have to lug or lift stoves and refrigerators that have no use or value.

If you are bringing your own stove and refrigerator from the city, make sure they will fit in the locations where you intend to place them. Make sure the proper gas and electric connections are available.

The kitchen should have a handy rear entry. This is probably the best place to use as a utility room to house your washer, dryer, and hot water heater. I believe that making a place for the woman of the house to do her washing and ironing upstairs is much more pleasant and labor-saving than to locate the laundry equipment in the basement where she will have to carry wash loads up and down stairs several times a week.

As long as water connections are handy, the rear entryway is also a convenient location for a lavatory or shower where the kids or the man who has just slopped the hogs can wash up and change without having to track through the house.

Before you leave this location, check the rear door. Is it warped? Does it open, close, and lock with ease? Does it have weatherstripping or is there a draft coming in because it's not snug?

Now that you've had a chance to look the whole house over and draw a floor plan, take a minute or two to figure out both the upstairs and downstairs traffic patterns. It is usually best when a main hallway connects with the rooms, providing easy entry. You don't want everyone to have to cut through the living room to get to the kitchen.

Electricity

Rural electrification has now spread all across the American countryside. And the success of the programs that brought electric power to our farms has to be counted among the great national achievements of this century. Electric power has helped make the life of the American farm family more comfortable and more productive than that of their counterparts anywhere else in the world.

But the job of electrifying our farms and farmhouses is a relatively recent one. Many homes were built years before their fuse boxes were installed. In a fair share of them, the wiring is inadequate or obsolete. As you inspect various farmsteads, looking for the one you will eventually buy, you are going to find quite a few with wiring systems incapable of handling all the various appliances we city folks have come to regard as necessities. Sometimes the wiring will not conform to modern safety codes and insurance requirements.

As you inspect each room of the house, try to visualize the lamps, clocks, radios, hi-fis, TV sets, etc. you might be using in that particular room. In bathrooms think of hairdryers, electric shavers, electric toothbrushes, etc. Try to judge whether the wiring already installed will meet these needs. In the kitchen, remember the mixers, toasters, disposal units, clocks, radios, pressure cookers, dishwashers, etc. Even if you don't have all these appliances right now, you may someday.

The guidelines furnished by the National Electrical Code can be helpful to you in checking out various home wiring systems. A copy of the Code can be purchased for $3.50 from The National Fire Protection Association, 60 Batterymarch St., Boston, Massachusetts 02110. If you want to save your money, ask your local power company for any free printed material they may have.

According to the Code, which suggests only minimum requirements, all outlets should be grounded and installed a foot and a half above the floor at a distance of no greater than 12 feet apart along a wall. Normal home needs will require a system with a capacity between 100 to 200 amperes. Keep in mind that the electric stove you've grown fond of and intend to bring with you when you move from the city will require a 220-volt circuit. Some of your other heavy-duty electric appliances such as clothes dryers, airconditioners, and heaters will have similar high-wattage requirements. The farmhouse may already have a 240-volt line, but if not, you'd better check with the nearest power company to find out how much it will cost you to have one brought in. If their quote is exorbitantly high, you may decide to dispose of your major electric appliances in the city where you can find ready buyers. When you move to the country, you can acquire replacements for them powered by gas.

If the seller already has some heavy-duty electric appliances, perhaps a workshop in the basement, you're probably safe in assuming the proper wiring has been installed. Then again, it never hurts to check the fuse box (or circuit-breaker panel). And if you don't know anything about electricity, it might be a good idea to call in someone who does to look over the system and see if it's okay.

Most houses that have been built or wired in the last twenty years or so will have the "three-wire" grounded circuits and the circuit-breaker panels required by some county and local ordinances. If a competent electrician or electrical inspector says that this house doesn't, then you had better be prepared to take on the cost of rewiring. A house in the country can be a long way from the nearest fire department. A safe and adequate wiring system is a lot cheaper than the cost of rebuilding after a fire.

There still are some farms with no off-farm power company near enough to provide low-cost electricity. If that's the situation where you plan to settle, and you like to watch the TV news before you turn in for the night, I'd suggest you find a farmstead that's closer in to town. If your mind is made up, you'll either have to do without the comfort of electrical appliances and tools or develop some sort of on-farm generator and power supply that can meet your needs at least part of the time. If it's impossible to get electric power, you might as well forget the telephone too—but maybe that's why you decided to move way out there in the first place?

I think you should know from the start that the cost of electric power in the country can be considerably higher than those monthly power company bills you've been complaining about in town. With the current inflationary economy, those costs are certain to continue rising. This move to the country can be a great opportunity for your family to get together and reassess the ways you may be able to conserve electricity.

You may be accustomed to heating and cooking with bottled gas. Or, you may be forced by circumstances to buy a rural home where bottled gas is cheaper, or the only fuel available. If that happens, have the tank and gas lines checked before you buy to see that they are in good working conditions and meet all safety requirements. If natural gas is available locally, call the gas company and have them give you an estimate of how much it will cost to have your new country home connected. If you give them a list of the gas appliances you intend to use, they should be able to give you an estimate of what your monthly bill will be under normal use. You can compare that estimate to bills that the current owner has been paying to the propane or butane company. In most cases, you will find that natural gas, supplied by the local gas company, is more dependable and economical in the long run.

The Heating System

When Ilene and I were looking for a lake cottage in northwestern Michigan a few years ago, we searched until we found one that had a fireplace in the master bedroom. I guess we thought it would be romantic to fall asleep by firelight. Well I don't know how romantic it was, but it sure was cold! On winter nights, when the fire was roaring away, the bedroom got so hot that I kicked off all the covers. Then in the wee hours of the morning, when the fire died out, in order to save ourselves from freezing to death we had to scramble to pile on the blankets and hug each other so as to generate a little body heat. Come to think of it, that *was* romantic!

The recounting of the above little true-life tidbit is my way of emphasizing how important it will be for you to find out as much as you can about the efficiency and dependability of the heating system in any home before you buy.

I suppose that it goes without saying that the best time to check out a house's heating system is in the dead of winter, "when the ground is covered up with snow an' icy winds begin to blow!" The only trouble with that idea is that most folks with "farm fever" go shopping for their country home in the spring or summer. So what! Ask the seller or realtor to turn the heating system on anyway. This will let you hear and feel how it works. A noisy, laboring furnace may require closer inspection by a heating expert.

Before you go house-hunting it should prove helpful to learn a little bit about the various types of home heating systems you are likely to run into. If the house does have one of those big old furnaces like the one I mentioned earlier, the system is probably conduction-type central heating. It may be either a *gravity-feed*, which operates on the theory that the hot air will rise through pipes or a central register and the cold air will descend through return vents located next to cold, exterior walls or it may be *forced-air conduction*, which uses a fan to circulate the warm air through the house. Conduction heating systems are the only kind that can be adapted to central air-conditioning. If that is your eventual plan, keep this fact in mind.

Another commonly used version of central heating is called *radiant heat*. With radiant systems, the furnace heats a boiler which sends steam or hot water through pipes and radiators in each room. Some of these systems can be very efficient. But some of the older steam-type radiant heating systems can be very noisy. Some of the hot water-type are slow to respond to the thermostat because it takes a long time for the water to be reheated after it has cooled. As a result, the radiator is either ice cold or hotter than a pistol!

The newer forced-water types provide more constant and even heat. The water is kept moving through the pipes by a circulation pump. The most modern heating systems often combine forced-air conduction with air-conditioning, or forced-hot-water heat and wafer-thin, finned radiators combined with air-circulation fans.

Other homes use old-fashioned or modern space heaters to keep the rooms warm and livable. The heaters are usually strategically placed around the house and are often expected to be complemented by the heat from fireplaces. The trouble with space heaters is that unless you have one in almost every room, there are bound to be some cold spots in the house. Unless the house is small and cosy, or located in a warm climate, space heaters are not as efficient as central heating systems.

I should mention that fireplaces can be made into efficient space heaters if a curved air-circulating wall is installed in the back of them which forces the heat back through the screen and into the room.

The fuel or energy source for the country home heating system is often decided for you by local availability and custom. However, available sources and costs can change dramatically for various reasons such as the recent oil embargo. If your heating system is compatible, you may want to consider one or more of these heat sources.

Wood

Wood is probably man's oldest fuel and is still being used extensively. Houses which rely on fireplaces, wood stoves, or furnaces are best heated with hardwoods, which burn more slowly and give off the greatest amount of heat. Some homeowners have special preferences for the kind of wood they use as fuel. Aromatic woods are popular. Usually though, the type of wood burned is determined by what is available for sale in an area. The amount of wood you will need to heat your home through a winter will also vary, depending on the size of the home and the severity of the weather. It's best to keep several cords on hand and know where you can replenish your supply on short notice after a cold snap. A cord of firewood is a stack four by four by eight feet. Prices for firewood vary greatly. If you are a do-it-yourselfer and have a good ax or chain saw, you may be able to get permission from a nearby state or national forest to cut as much as you want from dead or fallen trees. Or if your farm has its own woodlot, it can prove to be a source of free wood. Commercial prices vary according to the time of the year. A cord of commercial wood, sometimes oak but often pine, can range between $25 and $90. Buy in the late spring or summer when it's cheapest.

Coal

This was once the most popular source of fuel to heat our homes. Since World War II, however, the costs of mining coal increased to the point where other fuels were more economical. Smoke and fumes from coal burning came under attack from environmental groups. Many coal mines played out. Opponents to strip mining of coal have also been instrumental in limiting production. Americans found it cleaner, less work, and more economical to turn to alternative fuels to heat their homes. Some of those fuels are by-products of coal; namely, electricity, gas, and petroleum.

In certain areas where it is being mined, coal is still the most common fuel source for home heating. Types used can be either "hard coal"—called anthracite, which burns cleanly—or "soft," bituminous coal, which emits more smoke and gas as it burns and makes more ashes.

Make a mental note that ashes from wood fireplaces and coal stoves and furnaces can supply needed potash to your garden plot or compost pile. Ashes are also good for fill or to increase traction on slippery driveways and walks in winter.

Another form of coal is coke. This is coal that has been processed to remove petroleum or gas. It burns quickly, giving off a high heat leaving little or no ash.

If the house you are looking at has a coal furnace, you may also find an automatic, or semiautomatic, coal stoker which will greatly reduce the need for you to do a lot of shoveling and hand-firing. Stokers are usually not suited to the use of coke. They work best with smaller-sized coal called pea coal.

The United States happens to be sitting on the richest coal and hydrocarbon deposits of any nation in the world. With the continuing concern over the depletion of the world's oil resources, it's pretty likely that there will be an increasing use of coal, or some form of it, to heat our homes in the future.

Oil

Many farms use oil to heat their homes and barns. Oil burners are not very trouble-prone. They are automatic and need very little attention from the busy farmer. Oil heat can also be used in barns and other types of livestock housing.

Some oil burners, usually the gravity-flow type, vaporize the fuel before burning it. This may make it necessary for you to buy a better quality, more expensive grade of oil to operate them. Have a heating expert check out the system to see what type it is, recommend the proper fuel, and make any adjustments to improve its efficiency.

Because of the widespread use of oil as a heating fuel and a power source on farms and in other businesses important to the nation's well-being, the federal government is working to keep the price of this critical fuel stabilized. If the energy crunch worsens and the government introduces rigid control measures, you as a farmer may be entitled to special fuel allotments. I suggest you keep abreast of developments in this situation.

Gas Heat

As a fuel, gas is quite efficient and burns cleaner than coal or oil. The old-fashioned coal furnaces found in many farmhouses are easy to convert to gas. Gas burners and furnaces are automatic and need only minimal maintenance. Yearly inspection, adjustments, and minor repairs are often done free, or for a nominal service charge, by the local gas company.

It is somewhat difficult to compare the costs of heating a home with gas and heating it with other fuels because costs are usually based on local availability and tradition. Usually, gas heat costs more than oil and less than electricity. As with electricity, your overall costs can often be reduced if you use other gas-operated appliances such as stoves, clothes dryers, refrigerators, and air-conditioners. This will cause your bill to be figured on a lower rate structure.

Electric Heat

The chances are not very great that you will find many old farmhouses heated by electricity. Rural electrification and all-electric homes are two fairly recent developments. It's uncommon to find cheap power in farm areas. However, if you do come across a home with electric heating, you should be able to weigh its good points against its drawbacks.

Electric heat is the cleanest now available. In new homes, electric heaters can save a great deal of space as they don't require a chimney. They can be located on the first floor if the house has no basement. It can save you dollars if the heating and all the appliances in your home are electrical. This is especially true if you happen to reside in a rural area that's near a major power center, like the TVA.

The biggest drawback, as mentioned above, is that electric power is usually costly in rural areas. Another problem with electric heating or an all-electric home, is that a severe storm could knock the wires down and cut off your service.

You will need to have an electrician check the existing wiring in a house which you wish to convert to electric heating. An electric heater will require a circuit with 200-ampere load capacity.

Solar Heat

Because of the energy crisis, there is bound to be lots of experimentation with new types of home and farm heating and cooling systems. One of the newest systems being implemented is based on an adaptation of the old idea of converting solar energy. Solar-heated homes have been built in Europe and in some parts of this country where there is a year-round supply of sunlight. These houses may become commonplace before the end of the century.

Insulation and Fuel Conservation

Basic common sense and the desire to obtain the greatest possible fuel economy dictates that any farm or country home should be well insulated. Usually, these structures are less sheltered and more exposed to the elements than homes in the cities or suburbs. In addition to applying or installing insulation in the attic, basement, or crawl space, the air space between interior and exterior walls should be filled with insulation. If the insulation has been installed after the house was built, you may find that this was done by blowing rock wool or other loose types of insulation down into the wall spaces from the attic. Cracks in the exterior foundation, caused by faulty construction, should have been sealed to prevent heat loss even when there is no danger of water seepage. All doors and windows should fit snugly and have weatherstripping applied wherever necessary to limit air exchange. Storm doors and storm windows can also save heating dollars. In homes that have central air-conditioning, the storms can be left on during the summer to help keep the cold air inside the house.

If a house has been carefully safeguarded with such heat-conservation measures, you—as the new owner—can realize major savings in your monthly fuel bill.

Humidity

In all of my books I have stressed the importance of keeping the humidity in your home relatively high. Humidity is extremely important to the health of plants and people. Too often, we keep our houses very hot and dry in the winter. Dry heat withers the foliage of houseplants and the skin of human beings. So, again, try to keep the relative humidity in your house (whether it's located on a farm or in the city) at around 40 percent. This moist atmosphere will not only be more healthful but it will allow you to maintain your thermostat at a lower setting because it is *dryness* in the air that cools off your skin and makes you feel cold. Maintaining 40 percent humidity in your home will make you feel warmer and result in noticeable fuel cost savings.

Plumbing

In the city, you and your family may have taken the operation of your plumbing pretty much for granted. If something went wrong, you probably just called a plumber and had him come and fix it. Out in the country, things aren't quite so convenient. You will quickly learn that the plumbing and its good operating condition are critical to a healthful and successful farm and farm family. Remember, the nearest plumber is going to be you!

When you look at a farmhouse, try every fixture and piece of plumbing equipment you can in order to see that it works.

Turn on the faucets in one of the bathroom sinks. Does the water fill the basin quickly and evenly? Or slowly and in spurts? Is the hot water hot? Are the hot and cold water mixed by one faucet, or are the faucets the older, separate-line type? If they are the latter, you may want to change them in the near future.

Turn on the water to fill the bathtub upstairs. Check the rate of flow. Does the tub fill quickly? Now, flush the toilet while the tub is still filling. As the flush tank refills, see if the flow of water into the tub is interrupted or the refilling ability of the flush tank is significantly lessened. If so, and you have a large family, you are going to have water problems. It will prove very inconvenient if the water pressure is so low only one person at a time can use something tied into the system.

The Water Company

If the house's water supply is connected with a main from the county or local water company, ask them to send a man out to check the flow rate and water pressure. If the water bursts from the pipe and splashes everything in sight every time you turn on a faucet, the pressure may be too strong. If you can't get a plumber or a serviceman from the water company, you can probably buy a screw-on water pressure gauge at the nearest hardware store or plumbing supply. The pressure gauge should read somewhere between 40 and 70 pounds per square inch. Anything below 40, and the pressure could be too low. Any reading above 70 could mean that there is more pressure than the plumbing can handle. This can be corrected by having a pressure control gauge installed between the meter and where the incoming cold water line connects with your system.

Another problem, common where a farm has to rely on outside water, is that the town supply may not always be adequate to meet the needs of a growing community, industrial demands, *and* the needs of outlying farms. It may have been a good move when the current owner first hooked on to the main of a nearby town. But since then, the town may have outgrown its water supply. It may now be a large suburb. Schools and shopping centers and industrial plants may have been built on acreage that once was farmland. It could now be that so many demands are being made of the water system that at certain critical periods of the year, water is in short supply and local ordinances have been passed limiting its use. If the kind of farming you plan to do uses large amounts of water, you should check with the local water commission to make sure there will always be enough on hand to supply your needs. If not, this is not the farm for you. When you talk to the water commission or water company, find out if any improvements of the system are being planned which will make you subject to any special tax assessments.

If local or county ordinances require that the farm be connected to community water and sewage but the current owner has failed to comply, you will want to make certain he does so before you buy the farm and have to pay such costs or assessments.

Water Rights

Sometimes water is brought to a farm by privately held water companies. If you are considering buying a piece of land which is part of a larger farm, or a farmhouse that's on a parcel of farm property that is being broken up, be sure to check on this possibility. *You will want to be absolutely sure you are buying the water shares along with the property!* It sometimes happens that a farmer will have already sold the water shares before he sells the house and land. Or he may sell the shares to the buyer of the other portion of the property, leaving you at that man's mercy if you need more water than he wants to allot you. If the property is being offered without the water shares being included and you are still interested in the farm, make your offer with a contingency clause that gives you the right to cancel if you are unable to secure the water shares.

If the farm you are looking at has a water supply which comes from a reservoir off the property or a stream which runs through other land owned by neighbors in the community, check to see if there are any legal restrictions to the amounts you can pump from the source. Also find out if the source is adequate for your family and farm needs all year round, and especially during the times you will need the most water for your crops and livestock.

Wells

In the past twenty years, there has been a rapid spread of commercial and local government-sponsored water companies and sanitary districts into rural areas. In spite of this, a majority of the millions of Americans who live in the country still depend on wells for their individual water supply.

The costs of running mains into outlying areas and maintaining them properly can be enormous. Under a great many conditions, a well *can be* just as dependable, more sanitary, and less costly. I can remember, as a kid, being repelled by the taste of "city water." Water from our small town or from out of a well always seemed to taste "purer." Recent tests of the water supplies in Cincinnati and New Orleans, which found cancer-causing chemicals, seemed to confirm my preadolescent sagacity. Now I find out that "town water" and "well water" are even less to be trusted!

The 1969 Community Water Supply Study by the U.S. Health Service showed that "major deficiencies in water system facilities increased as the population served decreased." * Two USDA experts estimate that "about two-thirds of the private supplies have some serious defect." †

It shouldn't be necessary for me to tell you that contaminated water can be extremely dangerous to your family's health. Diseases like amebic dysentery, hepatitis, and typhoid fever are often traced to faulty wells and contaminated water supplies. In addition, poor quality water from wells that have not been maintained at proper sanitary standards can spoil your chances of success with crops or livestock. *Don't start your new life in the country with contaminated water!*

Ask the current owner if he keeps a log on his well showing when it is inspected, repaired, and tested for contamination. Too many farmers are nonchalant about their water. If the seller doesn't keep any record of his well maintenance, you should contact the county health department and have a sanitary engineer check it thoroughly to make sure it is in a safe location, in good repair, and free of contamination. If this inspection does not give the water supply a clean bill of health, do not buy the farm until the owner shows to your satisfaction that he has corrected all problems, or dug a new well.

Just as important as the sanitary quality of the water is the dependability of the well to deliver the amount of service you need. Sometimes wells which have provided ample water for

---

* 1971 *Yearbook of Agriculture,* "Water Systems & Sewage," Jones & Hady, p. 226.

† 1973 *Yearbook of Agriculture,* "Well Water Supplies: Getting the Best from Your Own System," Pomeroy & Jones, p. 146.

years—all the way back to colonial times—may furnish an inadequate supply when new farming techniques or modern plumbing make greater demands on the system. It may be that you will have to dig a new or deeper well. If you find this out before you buy, it may be that you can transfer this cost to the seller.

AMOUNT OF WATER NEEDED PER DAY *

Household of four persons	150 gallons
Horse	10 gallons
Steer or dry cow	12 gallons
Milk cow	25 gallons
Hog	2 gallons
Sheep	2 gallons
Hens (100)	8 gallons
Lawn and garden (100 sq. feet/when needed)	8 gallons

* USDA statistics.

It's been estimated that the average farm family needs about 2,000 gallons of water a day. If you are planning to farm land that needs irrigation or intend to run a big livestock operation, you will need far greater amounts. The U.S. Geological Survey has probably determined the groundwater potential in the area. For information, write: Water Resources, U.S. Geological Survey, Washington, D.C. 20242. The SCS survey mentioned earlier may include enough information for you to estimate the groundwater potential, or your state water resources control board may have this information or be able to tell you how to get it. If several experts say there isn't enough water to meet your farming needs, find another farm.

Shallow and Deep Wells

When I was a boy, living at my Grandma Putnam's, I can remember that we had two wells on the property. One was the original shallow *dug-well.* Grandfather Coolidge, who homesteaded the farm, had dug it himself—down about 30 to 35 feet! He had curbed it himself too, no small job. By the time I lived there, all the curbing bricks—from the surface down about 15 feet—had been plastered over with cement. And at ground level, there was a concrete sanitary seal over the top of the well. Perched on that concrete platform was an old-fashioned long-handled pump. When a city boy first moves to the country, he just loves to pump water! This "affair" with the pump usually lasts about a day.

The other well was deeper—how much I don't know. Uncle Bob said it was a *driven-well,* made by having increasingly longer sections of pipe slammed into the ground, pulled up, and slammed in again and again until the point passed the water table and struck good water.

That summer, there was a long, dry, hot spell. After a few weeks, even though I was doing my part to conserve water by not taking any baths (whenever I could get away with it), the wells went dry. First, the dug-well dried up. Then the driven-well. Finally, Grandma called in Burt Barron, the most respected well man for miles around. He decided to convert the driven-well into a deep, *drilled-well.*

By deep, I mean that Burt drilled down about 190 feet. Once the drill gets past the top of the water table, into the *zone of saturation,* the drilling ceases as soon as the well driller hits good quality water. The zone of saturation is a stratum of sand and loose rock or gravel into which the groundwater has percolated down. It collects there, filling the spaces between sand, soil, and rock particles. The thickness of this zone varies from location to location. It take a competent and experienced well man to judge if the spot where he is drilling is liable to end up with a "wet" or "dry" well and whether the quality of the water is likely to be better at a certain depth.

After the hole has been drilled, to a level where enough good water seems to be present, a pipe or casing is sunk to protect the walls from crumbling. Then, the well driller seals the annular space around the well casing with a grout made of neat cement and water. The casing can be made of several materials and in several ways, depending on the makeup of the zone of saturation. Often, it has a screen at the bottom to keep out sand. The casing should extend above the surface to keep out contaminating surface waters. A thick sanitary seal of concrete protects the casing for 15 to 20 feet.

Every well, no matter how well constructed, will eventually have some need of repairs. If proper periodic inspections are made, each spring and fall, it should be possible to spot potential problems before they damage the farm family's health or become too expensive to correct.

Unless you are buying land to build on, you will probably purchase an existing well with your farmhouse and property. To check out whether it contains an adequate supply of water, find out its depth, flow rate, condition of its pump, and if it has ever dried up in the past.

Watch out for Abandoned Wells

Real estate agents will sometimes try to sell you on the idea that an old, abandoned well can be repaired. Before you accept that pitch, have it thoroughly checked out. The USDA says that if a well is contaminated, it's usually safer to drill in a new location than to try and clean it and reseal it.

A deep well needs a good pumping system. If there is a windmill being used for that purpose on the farm, have it checked to see it's in good repair. It is probably more likely that the well will have an electric pump. This can usually deliver all the water you need, even at peak water-use periods, for a minimal cost. The pump will be controlled by a pressure switch and pressure tank in order to provide the water pressure you need for modern farm, garden, and household appliances and faucets.

If the well on the farmstead is what is commonly called a low-yield well, the water system may require intermediate storage tanks to provide a steady supply during peak-use periods. Such tanks should be checked for condition, capacity, and sanitation. If the tank or tanks are raised, examine the condition of the supporting wood or metal structure.

The Hot Water Heater

Sometimes older farmhouses have hot water heaters that are old, obsolete, and inadequate. Find out the hot water heater's age, capacity, and how quickly it will deliver hot water after it has been refilled. The tank should be large enough to handle peak hot water needs including laundry and bathing back to back.

Water Softeners

In many parts of the country there is a high content of limestone, iron, copper, and other minerals in the water. These minerals can be harmful to pipes, stain clothes in the wash, discolor porcelain fixtures, and make the water taste bitter. The installation of a water softener will remove most of the unwanted minerals. If a water softener has already been installed, check to see if it is automatic, semiautomatic, or manually rechargeable. The seller will be able to tell you how often the system needs to have the salt or softening agent replaced and be backwashed.

An Emergency Water Reserve

Because of the distance between some farms and the nearest fire department, it may be necessary to keep an emergency supply of water on hand to be used in case of fire. Sometimes farmers get busy and don't maintain these tanks as well as they should. Be on the lookout for leaks or lack of pressure.

The Septic System

When you live on country property, just as you will probably have an individual water supply you will also have an individual sewage system called a septic system. It's been said that it was the invention of the septic system that got rid of the outhouse and brought indoor plumbing to rural America. A typical example of such a system includes a poured concrete or metal tank to accumulate sewage. The human wastes are carried to the tank by water. These wastes have a high bacteria content and are capable of spreading dangerous diseases to people inhabiting the property so the tank must be sealed tightly. Inside, the solid wastes are decomposed by aerobic bacteria and microorganisms. As they break down, they are carried into the drainfield, which is two loops of perforated tile sloping away from the house in subsurface trenches. The drainage loops should be laid out in an area of approximately 2,500 square feet. The tiles are buried at a minimum depth prescribed by local or county health ordinances (usually at a depth of 18 inches under grass or ground cover where plant roots will absorb and use some effluent). In areas of severe winters, the tiles must be buried below the frost line and insulated on top with a layer of gravel. The effluent, containing liquid and organic waste that has been partially decomposed, drains into a permeable soil mixture (usually silt and gravel). It then percolates down through the subsoil under the drainfield until it has been recycled into clean, rich loam.

A septic system can only be installed where there is enough water pressure to flush it. Normally, a newly installed septic system will work properly for about a dozen years—if the sludge and slime is cleaned from the septic tank every few years. Find out how long ago the present owner had this done.

Before you buy a farmstead, have health authorities check the system to see that it meets minimum sanitation standards. This can be accomplished by testing the percolation ability of the soil. If the field is improperly located—too close to a pond, lake, or well (closer than 75 feet to a water source) or where the subsurface soil has lost its percolation power—it should be moved. You can have a percolation test made by sanitary engineers for a minimal fee. Also, if the septic drainfield is located less than 100 feet from the farmhouse, it should be moved. If the septic system needs moving for any of the above reasons, it should be done at the seller's expense.

Waste Disposals

The farmhouse may have a waste disposal or you may have intentions of installing one of these after you buy. There's no doubt, they do cut down your trips to the garbage can. If there is one already installed, observe it under normal operating conditions. Does it have a reset button in case of jamming? Does the drainpipe it feeds have a straight or only slightly sloped drop to where it joins the main drain for the house? In order to work properly, a waste disposal should not feed into a horizontally sloped drain. Find out how old the disposal is and compare this to the age of the septic system. It has been estimated that for each year a food and waste disposal is in operation, it will deduct up to a half-year from the life of the average septic system. In order to correct this problem, the USDA recommends installing septic tanks with 750- to 1,000-gallon capacity for homes with waste disposals. If there is no disposal in the house, you may want to think twice about installing one in a country home not connected to local or county sewers.

If an outdoor privy is still being used on the farmstead or country acres you are looking at, make sure it has been designed to meet local sanitation and health regulations. Your county agent or health commission can tell you what these standards are. Also, check to see that waste holes beneath such privies are not so full that outhouses need to be moved very soon after you move in. Check to see that there is good air circulation inside. Is it comfortable in both summer and winter? In areas where the winters are cold, some outhouses have gas or electric space heaters. If so, make sure they operate properly. Is there an old mail-order catalogue?

How Do You Like the Place So Far?

Now that you've inspected the house and its facilities, you probably have a good idea if this is the farm for you. I'm sure you will find some of the drawbacks I've mentioned in this section. But you may have found a greater number of pluses which make the house and farm appeal to you as the place you would like to live. You should remember though, if the house needs remodeling or major repairs, you may have to put some of these off until more important capital expenditures are made. The farmhouse isn't a revenue-producing part of your country company, so don't plan on pouring your money into it first. Even if you like the house, before making an offer, I suggest you take a look at the other buildings and features on the farmstead. Also, you will want to assess the quality and condition of any livestock, machinery, seed, or feed being offered along with the house and property. In the meantime, fill in the farm scorecard in Section Two with the information you have accumulated up to now.

The USDA says that if you have to spend more than 20 percent of the amount of capital you have on remodeling and improvements to the property, you are not putting your money where it belongs. They say that 80 percent of your chances of making a good living on country land depends on you, the fertility of the soil, and the land's capability to support your type of farming. While the topics covered in this and the next section are important, they should not amount to more than 20 percent of your reason for buying a farm.

Section Four

Farmstead Features, Livestock, Seed, and Machinery

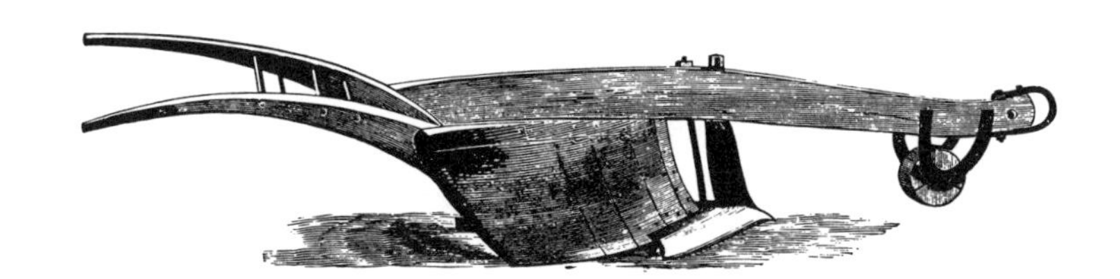

Now you can go ahead and evaluate the remaining features of the farmstead which may have an effect on your decision to buy or not to buy. Among these are: buildings, landscaping, improvements, livestock, seed, farm machinery, and the overall design of the entire farmstead and farm.

You should have already drawn a floor plan of the farmhouse as you took your "guided tour" in Section Three. This will prove invaluable in helping you plan your family room allotments and furniture arrangements. Now I suggest you draw a similar plan or map of the entire farmstead lot. This map should show the location of the farmhouse, septic system, well, and other important features like: the farm buildings, landscaping, patio, pool, farm garden, orchard, fencing, driveways, and access road.

As you inspect these various features more closely, the farmstead map will be useful in helping you decide if the layout has been designed with an eye to location, sanitation, traffic patterns, comfort, and appearance. USDA experts suggest that the farmstead lot be at least 1 acre; 2 to 5 acres if there is a barn, corrals, and livestock pens. Perhaps another ½ to 1½ acres if there is a family farm garden and a home orchard.

Sometimes the farmstead has been located where there is the best view of the countryside. More often, other considerations have influenced the builder to select the site where it is now located. Among these are the availability of water, shelter from harsh weather and prevailing winds. Proximity to roads, utilities, or neighbors are other considerations. Other times a location which looks good is compromised by limiting factors like poor soil percolation for sewage or lack of good air circulation, which is a must for raising healthy livestock.

Look at your farmstead layout objectively and try to decide if this is the best possible location on the entire property for the place where you will live and spend most of your time. If it is not, and if many of the buildings—including the farmhouse—are obsolete, you may want to buy another part of the property without the house and the farm buildings. Then you can build a farmstead where the location is more suitable.

If the farmstead was built by a man who liked the view or loved his privacy—15 miles back up in the hills—think about getting your kids off to school every morning! Also, think about driving down to the road to pick up your mail every day; the possibility of being cut off during the winter by heavy snows which block your driveways and access roads. Think about the cost of bringing in needed new utility lines, trucks to transport your farm products, or the time lost bringing the doctor up there when an emergency arises. Sometimes privacy and a nice view grow pale beside other, more practical considerations!

Were You Born in a Barn?

Back in the old country, farmers used to have their cows, pigs, and chickens in to sleep with them at night. Experts will tell you this was because they were less secure about their prized possessions than we Americans are. I'm not so sure about that. What about all those folks who bed down with their cars in a downstairs garage? Maybe it's not so much that we dislike animals in the house as it is that our conception of prized possessions has just changed over the years.

Some people are never satisfied. Not happy with having chased the animals out of the farmhouse, they've followed them out to the barn and kicked them out of there! That's right, since World War II, thousands of city folks bought old barns, remodeled them, and turned them into houses. It became so popular to do this up in New England that, by now, there must be lots of kids running around who would have to answer with a hearty yes when someone asks if they were born in a barn. If you have such a remodeling project in mind for your country place, I suggest you get an architect or contractor who specializes in such work to inspect the building to see if your plans are economically feasible.

The Farm Buildings and Livestock Pens

If you are fairly certain you don't want to live in the barn, you will still want to carefully inspect it and all the other buildings and enclosures on the farmstead, starting with the garage. Farmstead planners estimate that a farmer spends a minimum of a third of his time working in and around these buildings. Often, very much more.

The Garage

Just as in the city, the farmstead garage is likely to be overlooked in a prospective buyer's tour of inspection. This is a mistake because protective housing for autos and trucks is critical to a successful farm operation. And repairs to such structures can be very costly. The garage should be at least large enough to accommodate your family car or station wagon. A two- or three-car garage can be very useful if you have a large family or plan on having a pickup truck. If the farm is located where the winters are severe, it will be wise to park these vehicles out of the weather each night. Make sure the garage is weather-tight. Is the exterior in need of paint? Do the windows and doors open and close properly, even in freezing weather? Is there an electric automatic garage door opener? Is the roof in good condition? Does the concrete floor have a drain? Are there any cracks in it which indicate poor footing and settling? Is there room in the garage for a workbench or storage for auto repair tools? Can it be used to house lawn mowers and other garden tools? If other auto housing is available, could the garage be converted into a stable or housing for small farm animals?

If the farmhouse is of recent vintage and the garage is attached to it, make doubly certain that there is no way for auto exhaust fumes to enter the living quarters.

The Barn

If you are at all serious about farming, the barn will most likely become the most important and most-used building on the farmstead, excluding the farmhouse. Many older barns are obsolete. If you plan to start a dairy operation, you will have to get an inspector to go over the barn to see that it meets all the state and local health regulations affecting milk production.

Because of the rigid sanitation regulations for dairy barns, cows, and equipment, many small-scale dairymen were forced out of the business in the past fifteen to twenty years. This was because of the high costs of converting older barns into the ultrasterile cow housing and milking parlors of today.

Unless you plan to be a commercial dairyman, it will be unnecessary for you to have facilities with concrete flooring, modern cream separators, and milking equipment. If you only intend to produce enough milk for your own family, it will still be wise to be sure that the barn is clean. If you follow reasonable habits of cleanliness, like washing your hands and the cow's udders before and after milking, your milk will be just as clean as if it was piped through all that fancy equipment.

Don't jump to sign the sales contract just because a farm happens to have a beautiful barn. Elaborate and specialized dairy barns may be attractive to the untutored city dweller, and the seller may be justifiably proud of his, but they can rarely be considered worth their fancy price tags unless you know the dairy business and intend to be a large-scale, full-time operator for many years to come.

The barn should be located downwind of the farmhouse so the strong smells of animal manures aren't constantly being served up with the family meals in the kitchen and dining room. However, it is usually best if the barn isn't any farther than about a hundred yards from the house. If it is, you may find yourself walking an extra two to five miles a week just going back and forth.

Barns are usually built to certain common specifications. As a result, most of the barns you will encounter on your farm-buying inspections will be pretty much alike. Most barns are rectangular with the short ends usually facing north and south. Inside, the stalls are built along both of the long sides of the rectangle so that the farm animals will get the morning or afternoon sun. If the original owner and builder just had a small dairy operation, he may have situated the barn the other way. This allowed him to place the stalls for his smaller dairy herd along the south side of the barn.

Stalls for dairy cows are usually built with the specific type of cows who will occupy them in mind. If you are planning to raise replacement heifers, you will require a stall size that is much smaller than the 6-by-4-foot stalls installed by the original owner who had a herd of Holsteins. The length of the barn will relate to the number of stalls. Each stall is usually 3.5 to 4 feet wide.

While inspecting the barn, ask these questions: Is it heated? Has insulation been installed? What about electricity? What materials is the barn constructed of? Do the animals face in or out when they are in their stalls? (Outside-facing stalls are claimed to be best.) Is it easy to feed the animals and to remove the manure? Is it easy to separate the milk herd from the other animals being housed there? Is there a concrete floor? Is the barn damp? Is there good ventilation in the eaves of the building? What about electric ventilation fans? If these have been installed, are they in good working order? Is there an ample and dependable supply of water to the barn? Is there some sort of automatic feeding system? An attached milking parlor?

Sometimes, you will find an attached veterinary stall. This can be very useful if it becomes necessary to isolate an animal during calving or periods of sickness.

These are just some of the factors you will want to consider when evaluating a dairy barn. If the barn is one of the newer total-confinement type structures used in beef operations, it will have another, entirely different set of evaluators. You can work out a checklist of these factors by consulting other farmers who run total-confinement feeder operations, or you can consult your county agent or state extension service.

Horse Barns

It has been estimated that there will be a horse population of 14 million in the United States by the end of this decade. Because of this enormous surge in horse ownership, you may have become interested in operating a stable, boarding facility, or just having a few horses of your own.

Box stalls in horse barns are normally 12 feet by 10 feet by 8 feet high. The partitions between stalls should be of solid construction and approximately 5 feet high with steel bars or grates completing the partitions at 7 feet.

Horses for show should be kept in heated barns. In the South or Southwest, they may be of the shed or open-front type. Horse barns should be equipped with a tack room and have ample storage space for hay.

Recent trends in barn construction show that the modern farmer leans to one-story open-span barns usually of the rigid frame-type construction. These newer barns seem more suited to most types of modern agriculture.

The Hog House

Whether you plan to raise one or two pigs a year to provide your family with pork chops and bacon or you intend to make swine production a major part of your farm enterprise, you will want to inspect the farmstead buildings being used for hog housing very thoroughly.

First, look at the location where the hog house now stands. Is it downwind of the farmhouse and of sufficient distance to keep odors away? Is there enough water? Is there good drainage? Are the buildings separated enough to insure against overcrowding and disease? Is there easy access to feed storage? Is the setup such that daily feeding chores can be completed quickly and easily? Is there room enough for possible enlargement of the facilities if you should want to expand? Is there ample protection from the wind and weather?

Now look at the hog houses. Are there separate, colony-type houses for each sow and her litter (the clean ground system), or are there one or two central hog houses of the total-confinement type? Whichever type it is, does it fit your swine production plans?

Are the colony-type houses warm, dry, and disease-free? Are the pens that surround them clean? Is the soil disease-free? Is there an availability of both sunlight and shade? If the location is hot, is there a hog wallow for the pigs to cool off? Are the houses and pens near enough to pasture? Are the pens and pasture properly fenced for pigs—with an electric "hot line" placed 3 inches above the bottom to discourage rooting?

If the hog house is the total-confinement type, does it provide for climate control, proper sanitation, automatic feeding, efficient manure and liquid waste removal? Does it have a concrete floor which can be easily cleaned? Does the sunlight enter the hog house? Is there proper ventilation? Is the building in good condition or in need of repair?

A shed or building which is now being put to some other use may be easily converted into a hog house if it is easy to clean and meets most of the requirements mentioned above. Or if there is ample room on the farm lot, you can easily build your own simple setup for one feeder-pig. This would include a small house (approximately 10 feet by 5 feet)—of wood, concrete block, or some prefabricated materials—a pen, water, and creep-type feeder.

Ask your county agent for advice on repairing older hog houses or converting unused structures into housing for swine. Write your state ag-extension service for bulletins and information on new hog house construction.

Poultry Housing

Farmstead poultry houses are usually located where there is a sunny southern exposure and protection from prevailing winds. If the chickens are to be a sideline, or "egg money" operation, tended by the farm wife, the hen house and chicken yard should not be too far from the house. Proximity to the farmhouse will also make the housing accessible to water and electricity for light and heat.

The chicken house should be built on land that has good air circulation and water drainage. If you intend to practice different types or phases of poultry management at one time—brooding, rearing, breeding, laying—you will find there are different housing requirements for each. For the egg-producing phase, you will see floor-managed housing or caged housing. The same all-purpose hen house or chicken coop can be used if you intend to raise chickens in each phase sequentially. However, in this latter instance, you will necessarily have to make some adaptations in the housing and equipment as your flock begins each new phase.

Consult the seller about the chicken housing he has installed. Watch him as he tends the flock to see if the housing is convenient for the operator to do his chores. For the birds, it should be comfortable, clean, dry, and disease-free. Check with other poultry producers, the county agent, and the state extension service for more detailed information on poultry housing. You can write to the Superintendent of Documents, U.S. Government Printing Office, Washington, D.C. 20402, for the bulletin entitled "Housing & Equipment for Laying Hens for Loose Housing," (Catalog # A1.38:72813 S/N 0100–0428; it costs twenty-five cents.) Or a plan for a modern floor-managed laying house (2,000 layers) can be obtained from the Michigan Ag-Extension Service, Michigan State University, East Lansing, Michigan (ask for Plan # 727-C1-87). For the beginner, who wants a family-type operation for food and eggs at home, I recommend reading *Starting Right with Poultry*, by G. T. Klein. (If you can't get this helpful book at your library, send $3 plus postage to Garden Way Publishing, Charlotte, Vermont 05445.)

Housing for Small Farm Animals

As you inspect the various farmstead buildings and examine the layout you may have drawn, be on the lookout for places that may be good locations or structures for housing small farm animals. Carefully inspect any small animal housing that has already been constructed, whether it is presently in use or not.

Sheep Sheds

Sheep are usually pastured in some protected area during winter like a cornfield or woodlot. Where the weather is severe, they may be kept in sheds of the pole-type construction (described in Section Two). These sheds may have doors, but they are usually kept open at the side away from the wind. The sheds should be adjacent to feed-lot exercise pens or pasture so the sheep can be let out during periods of nice weather. Ewes about to lamb during the winter should be placed in warm, dry sheds with plenty of straw bedding. Some sheep operations may also have colony-type brooders in the lambing pens. If the flock is large, temporary lambing pens are erected in the pasture to provide weather protection.

When inspecting sheep sheds, pay special attention to the pole supports and the conditions of the joints where the pole or wall connects to the roof. Additional information about sheep housing can be obtained by contacting your state extension service, USDA bulletins, or the book *Sheep Science* by W. G. Kammlade (New York, J. B. Lippincott Co., 1957).

Goat Housing

If there is no structure on the farm that has been specifically built as goat housing, one or two animals can be temporarily housed in a small shed or empty garage. There should be an ample supply of water nearby and a place to store feed out of the weather. Goat housing should be constructed of sturdy materials as they love to chew and pick at insulation batting and will tear up thin plywood as though it were lettuce! Concrete floors are preferable to those constructed of wood or dirt as they are easier to keep clean. As with most small animal housing, it is important that the stable is warm, dry, and clean. Cleaning should be done on a regular schedule about twice a year—unless the weather is mild all year round. When you enter the goat shed, your nose will tell you if the current manager keeps it clean! If you intend to house several goats together, be sure that the partitions between stalls are open enough to permit them to see each other but high enough to prevent them from climbing over to get at one another. A good reference source on goat management is *Goat Husbandry* by David Mackenzie (Levittown, Long Island, Transatlantic, 1961).

Rabbitries and Rabbit Hutches

Rabbits are very easy to house. A good location for an outside hutch is in almost any sheltered spot where there is good air circulation, shelter from winds and extreme cold weather, and shade in the summertime. The location should provide some privacy and be where it is easy to clean under the cages or maintain a garden compost heap.

If the seller has a rabbitry, it may consist of several hutches or a completely climate-controlled building with cages (3 cubic feet) hung about 40 inches off the floor. Underneath the cages he raises earthworms in specially constructed metal bins. The worms grow fat off a combination of rabbit droppings, cottonseed meal, and compost. The compost is changed regularly so that the farm garden gets what is "ready" and the worms get a new dinner.

Information on rabbit raising can be obtained from the Superintendent of Documents, U.S. Government Printing Office, Washington, D.C. 20402. Ask for "Selecting & Raising Rabbits" (Cat. # A 1.75:358, S/N 0100-2640, fifteen cents) and "Commercial Rabbit Raising" (Cat. # A 1.76:30912, S/N 0100-1376, thirty-five cents).

Housing for other small farm animals with profit-making potential for the farm operator may or may not have already been installed on the farmstead. If it has, no doubt the seller will want to include these structures in the sale and use them to jack up his price. If so, you will want to use good common sense to inspect them carefully and see if they have been constructed and maintained properly. It could be that he has been making a profit by raising something you are not interested in having around. For example, there's a man in New Mexico who collects and sells scorpions, rattlesnakes, and exotic spiders. You or your wife might not want to copy his way of making a living if you bought his property. Therefore, the fact that he has invested considerable amounts of money in a structure to house those little dandies—including a laboratory where he milks them—is not necessarily something you will be willing to pay extra to get along with his nice little place in the country. Of course, this thought applies to machinery and livestock as well as buildings, *don't pay extra for something you don't want in the first place!*

If you do intend to raise certain small animals, and the present owner does not have housing for them now, look for buildings you might be able to convert. Keep in mind, though, that buildings which were built with one purpose in mind—while they may be converted—are not always the most efficient structures when used for another purpose. This is another plus or minus you must figure into your final assessment.

In order to be able to judge which structures are the most efficient for the type of farming you intend to follow, you should obtain as much information as you possibly can from other successful producers in that line. Also, take advantage of the many books that are available dealing with management and production techniques on almost everything from beekeeping to mushroom growing. These will give you added information on the types of housing and buildings best suited to the specific farm enterprises you plan to begin.

Other Necessary Farm Structures

In addition to the types of buildings I have just been discussing, the farm you are looking at will, no doubt, have other necessary structures which are used to store grain, process feed, and house machinery. These too should be carefully considered in terms of how they will fit into your planned farming operation.

Grain Storage, Drying, and Handling Facilities

Don't be a dummy and pass by the grain storage, drying, and feed-handling facilities when you inspect a prospective farm. These structures should never be taken for granted. Ask the owner-seller to give you a step-by-step explanation of how the facilities are used in his operation. See if the setup makes sense. Where these structures are located on the farmstead, how they are constructed and how well they work will either save or cost you a great deal of time, labor, and money in running your farm.

Old-fashioned corncribs and granaries have pretty much disappeared from most farms. The storage and drying bins that have replaced them are typically of two types. The first type is cylindrical, usually made of corrugated steel with a metal roof that slants up to a central outlet vent in the center. Other vents, which can be opened or closed, are also located in the roof. The interior of the bin has a false floor and air is forced up through the floor, on through the grain, and out through the roof saturated with moisture. The second type of bin is the square or rectangular type. In addition to the difference in exterior shape, these bins usually do not have the interior false floor. Instead, a series of air ducts has been installed which carry the forced air to the grain. The advantages of this second type of bin construction is that it is cheaper and the ducts can be removed and cleaned. You should inspect the bins to see that they are in good repair and the fan motors to see that they are in good condition and operating properly.

The capacity of the storage bins will vary according to the size of the farm and the amount of grain annually produced. In *Prac-*

tical Farm Buildings, Dr. James Boyd suggests that it is probably best to have at least two bins of unequal capacity in order to handle grains of different types, qualities, or harvest volume.

In recent years the drying systems have tended to change from unheated air systems to heated air systems. You should find out which of the two is being used on these facilities. If a heated air system is being used, you should determine the type of fuel used, the condition, capacity, and efficiency of the system. Sometimes aerators are installed with these systems to keep the drying grain at a constant moisture-holding level. These should also be checked.

Suggested dryer and grain-feed center specifications and layouts can be obtained from professional farmstead planners like the Butler Mfg. Co. or The Midwest Plan Service, Ames, Iowa. Or you can contact the USDA and state extension services for bulletins which give types of construction, ventilation, foundation, and pest-protection details and information.

As you evaluate a prospective farm's grain and feed-handling facilities, most experts say you should look for bin fill and drying facilities which are centrally located on the farmstead lot. They should be easily accessible to: 1) trucks and wagons from the fields; 2) the livestock areas for feeding; 3) transport vehicles taking the grain from the farm to markets. If the feed storage is centrally located, it may be possible to automate daily livestock feeding chores. Even if automated feeding is impractical or too costly, a centrally located feed center will enable you to prepare all the feed for various farm animals at one time and save many steps.

Silos

Silos are the storage structures which turn chopped green corn or forage crops into silage. Silage is the livestock food resulting from a fermentation process which takes place when these cut and chopped green crops are stored in a relatively air-free closed space. Often, molasses, uric acid, or other additives are mixed with the crops to speed up the fermentation process. The finished silage looks brown and ugly, but it must taste good because cows and hogs go bananas for it. There are basically three types of silos: trench, bunker, and upright. The more common upright silos can be constructed of wood, concrete, metal, or tile blocks. Because the green material usually has a high percentage of moisture in it when it is put in the silo, it exerts a great deal of pressure against the walls and foundation during the fermentation process. It is important that the building be strong and able to support this tremendous pressure and the weight of the stored crops. The older wood silos may need seasonal repairs because of the constant process of swelling when full and shrinking when empty. Inspect these silos carefully.

The newer oxygen-free silos have become increasingly popular in recent years and you may be lucky enough to find that the farm you are inspecting has one or more. In these silos, crops with a lower level of moisture can be stored and the resulting silage will be both higher in food value and will have a considerable higher cash value per ton.

Vegetable and Fruit Storage Facilities

Unless you are planning to go into fruit or vegetable farming in a big way, you will probably use commercial off-farm storage for these highly perishable crops. If you are interested in raising just enough fruits and vegetables to feed your family, these can be stored in a cold cellar. Sometimes cold cellars are freestanding (or buried) bunkers dug out of the side of a hill. More often they are separated unheated rooms attached to the basement of the farmhouse. However, it may be that the farm has a more modern fruit and vegetable storage facility. This should be evaluated as part of your farmstead survey.

Most of the fruits and vegetables you produce will have to be cool-stored until shipping time. This will slow down the decaying process which begins right after harvest. Check out any refrigeration, hydro-cooling, or vacuum-cooling facilities. Certain fruits like prunes, raisins, apricots, peaches, and apples have been sun-dried for generations by growers with specialized markets. A recent innovation in the prune industry has been the introduction of a dehydration-rehydration system for storage facilities.

Climate-controlled storage for apples and other fruit may involve expensive special equipment, but will increase the storage life and value of your crops immensely. You should get an expert to appraise the value of all sophisticated storage equipment.

Machinery Sheds and Storage Buildings

Wherever farm machinery and the gasoline to run farm equipment are kept, there is always a potential fire hazard. For this reason, you should make sure that the housing for such machinery be located at some distance and downwind from the farmhouse. Open-sided machinery sheds should be constructed with their backs to the prevailing wind in order to keep them free of snow and protected from freezing rain or sleet. If possible, they should also be turned away from the view of passersby on the main road fronting the farm.

Milk Houses

Most sanitary codes require that the milk house should be at least 20 feet from the barn. On many farms the milk house is located in a grassy area about halfway between the barn and the house. The USDA recommends that the building (100 or 200 square feet depending on the size of the herd) should have concrete flooring with a drain for easy cleaning and a tile or painted interior. Your local milk producers' or dairy producers' association will furnish details on health and sanitation requirements.

"There'll Be Some Changes Made!"

If after a thorough inspection of an otherwise suitable farm you conclude that the location and condition of the farmstead buildings are poor, you should give some serious thought to how you would change the setup. Or, if you do decide to buy the property, whether it would be wiser to tear down the old buildings and build anew.

If you are one of those people who has a good eye for remodeling possibilities, you may have already formulated several alternative plans for changing the layout and functions of several buildings on the farmstead. Perhaps you want to enlarge the barn to make room for more animals? Perhaps you want to turn it into a huge barracks-type chicken house? If the foundation is good but the building is in poor condition, you may want to tear it down and put up a modern one-story barn which will be more suited to the housing and feed-storage methods and the machinery used today. Whatever your remodeling plans, I suggest you keep in mind that a good third of all your on-farm work will be done in the area around the livestock-housing facilities or feed center. Converting older buildings which have been built for one purpose to make them functional in a new way always raises the possibility that you won't be happy with the conversion after it has been completed. Plan carefully; after you've spent a lot of money remodeling, you're the one who will have to live with the results. If you plan to tear down certain buildings which have no immediate apparent function in your intended farm operation, go slow. Sometimes conversion possibilities occur after it's too late.

If you are planning to produce the same kinds of farm products as the farm family living here now, take the time to follow them around as they do their chores. As you watch, ask yourself how you can do these daily jobs faster and better while expending less energy. You may discover that some of the chores don't really need doing at all. Sometimes farmers fall into the habit of doing a whole slew of unnecessary tasks. Often these little "make-work" jobs can be combined with other tasks, saving steps, or they can be eliminated altogether. Experts say you should work out a circular traffic pattern for doing the chores. That way, it won't become necessary to retrace your steps over and over again.

It may be that if you had your druthers you'd build a whole new farmstead. In order to find out whether this would be economically feasible, I suggest you consult with one or more of the firms which specialize in building modern farmsteads and which have farmstead planning services.

The Butler Manufacturing Company at 7400 East Thirteenth Street, Kansas City, Missouri 64126, is one of the largest producers of rigid frame, open-span farm buildings. They will be happy to send you various booklets and bulletins which indicate the various farm buildings and services they sell. Also available is a complete farmstead planning service from Butler. Another source is The Midwest Plan Service, Ames, Iowa. If you have to build, the use of services like these which do not commit you to buy a certain company's products will be helpful to you in obtaining competitive bids.

An excellent handbook which outlines the principles of planning farm buildings and details the specifics of construction is *Practical Farm Buildings,* by James S. Boyd (Interstate Printers & Publishers, Inc., Danville, Illinois, 1973).

The Farmstead Landscape

I've been told that every landscape has its own special kind of attraction and beauty. My friend, Col. Al Worden, the former astronaut, says that even the lunar landscape has a compelling and lonely loveliness that strikes a poetic chord in those who see it. Maybe that's so, but I'd hate to have a cheese and dairy farm up there without any trees or shrubs to help keep the moondust out of my green milk!

Jimmy Dean, the silver-tongued singer and praiser of pork sausage, says the moon kinda reminds him of the treeless family farm where he grew up back in East Texas. "More dust and dirt blew in through the keyhole than our whole family could shovel out the windows!" Now holding the dust down and helping to retard soil erosion isn't the only plus provided by good landscaping on a farm, but it is an important one. A farmstead that has a good landscaping plan will be well protected from the seasonal batterings of wind, snow, and rain. It will be warmer in the winter and cooler in the summer. It will be the kind of attractive setting where you and your family will want to settle and make your home.

Maybe, up to now, you've been so busy stabbing studding in search of dry rot, or measuring the machinery shed to see if it has ample floor space, that you haven't had a chance to take a good look at the trees, shrubs, and plantings that make up the farmstead landscaping! Well, here's your chance to give it a good once-over.

Begin by drawing in all the walks and driveways on your farmstead map. Now, add all the trees, shrubs, and plantings that are growing on the property. Next go for a walk around the place to inspect them more closely.

Shade Trees

Out here in the country there should be room to grow some of those huge old-fashioned shade trees that are too cramped for space on our modern city or suburban lots. An oak, maple, elm, or walnut needs room to spread out and show off its foliage. On a farm, these trees can add a feeling of stateliness to the house and grounds. On a sunny summer's day, it's great to lie down in a hammock under the friendly branches of one of these giants and watch them try to catch the breeze with their leaves.

As I said, these trees need room. Make sure they have not been planted too close to the farmhouse or to utility wires where ice-laden or windblown limbs will cause damage. Certain trees, like the silver maple, have brittle limbs that are especially subject to storm damage.

Other species, like the weeping willow and the tree of heaven, can develop roots which have a nasty habit of tearing up driveways, septic tank tiles, and sidewalks. If such trees are on the farmstead, see that they have been planted well away from the house and septic drainfield. On the other hand, weeping willows can be used effectively to dry up spots on the property that get soggy in the spring or fall rainy season.

One tree that you are likely to find on older farms is the great old American elm. Sadly, it is very susceptible to the dreaded Dutch elm disease. If you find several American elms on the property, be wary. If one of them should happen to be infected with the disease, you might end up having to pay hundreds of dollars to have the whole stand cut down.

Look for soil compaction or sogginess around trees. These are danger signals. Tree roots get most of their nourishment from the top few inches of the soil around them. If this is compacted or soggy, the roots may not be able to get enough air or food.

Look for damage to limbs or trunks caused by wind or storms. Also be on the lookout for signs that a crazy amateur tree surgeon has hacked away important limbs or branches. A tree may never be able to recover from this kind of unkind treatment.

Every professional landscaper has his own particular favorite trees which he will be happy to recommend to you. However, if you are looking at a well-established farmstead, you will probably have to live with the trees you find growing there, at least for the time being. It is always good to find out as much about these new big-shouldered companions as you can. To obtain this information I suggest you write to the Superintendent of Documents, Washington, D.C. 20402, and ask for "Shade Trees for the Home," *Agriculture Handbook* #425 (seventy-five cents) or contact the nearest arboretum in your state.

Trees and Shrubs as Hedges

Trees and shrubs can be used as hedgerows and to serve many useful functions and solve many problems on a farm. They may rim the outside boundary of the farmstead property providing an attractive setting and needed privacy for the farm home. They may be used to block off views of unsightly machinery sheds or waste disposal ponds. They may be employed as snow fences or windbreaks. They may be planted around a pool or patio to provide seclusion from passersby or nosy neighbors.

Sometimes the idea of perimeter hedges improperly carried out can have its drawbacks. Check to make sure that the farm wife has a clear view of the fields from her kitchen window. This will make it possible for her to signal when meals are ready and to spot any accidents should they occur. Perimeter hedges too close to the house can block out the sunlight all year round, keeping your home cold and damp. Tall hedges too close to the driveway can block off your view of visitors until they are all the way up to the door. Perimeter hedges too close to where the drive and the access or frontage road meet will block off the view of vehicles leaving and entering the farm lot, making your entrance a safety hazard. Keep these things in mind when you inspect.

These are some of the trees and shrubs that make excellent hedges.

Arbor vitae	Lombardy poplar
European beech	Honeylocust
Black thorn	Lilac
Hemlock	Oleander
Japanese yew	English hawthorn
European hornbeam	Osage orange

Small Fruits for Country Pleasures

Small fruit plantings are often used as fences on farms. Delicious currants, gooseberries, raspberries, dewberries, and loganberries are all commonly used to enclose work or play areas and to define property lines. Grape arbors, neatly trained, are often used as partitions. On my Grandma Putt's place, arbors were used in this way to screen off the work area where she hung the wash out to dry. If any of these fruits are used on the farms you inspect, see if they have been properly trained and pruned.

Bramble berries, especially, require a severe pruning each year or they will become unruly.

On your tour, you should also look for low, boggy, or "cold" spots where you might want to plant cranberries or blueberries. Small fruits in your farm landscape will make living on the farm a real pleasure. If you've never had a piece of gooseberry pie, you've missed one of the great country taste treats!

Espaliering Is a Sign of Loving Care

One sure sign that the seller has given loving attention to his farmstead landscape will be the presence of attractive patterned plantings called espaliers. Dwarf peaches, apricots, figs, and nectarines are just a few of the fruits which respond very well to this type of training. In *Talk to Your Plants,* I have outlined step-by-step instructions on how to plant and train some typical espaliers. Even if there are none there now, on your walk around the farm lot, look for places where you might try your hand at creating these artful and fun fruit-giving hedges.

Nuts!

Nut trees also will put some "meats" on the bare bones of any landscape plan. One of my secret dreams is to someday be able to sit under my own black walnut tree eating homemade vanilla ice cream with a homemade black walnut fudge topping. Sounds great, doesn't it?

But there are a few drawbacks to some of the nut trees being included in a landscape plan. One is that walnuts, hickories, pecans, butternuts, and most other popular nuts require the company of a member of the opposite sex in order to produce a crop. And, to add to your space problems, some of our nutty friends grow very big. My Grandma Putt had a couple of black walnut trees that scraped the sky at one hundred and thirty to one hundred and fifty feet! So space will be an important consideration. If there are a couple of young nut trees on the lot, make sure the seller has included them in the deal.

Of course, you can always grow some peanuts in your farm garden if you live far enough south. Peanuts only grow about a foot to a foot and a half high, and maybe, if you're lucky, someone will make some old-fashioned peanut brittle. Here's a list of some friends of mine who will be delighted to add a little nuttiness to your life on the farm:

Almond	Pecan
Beechnut	Pistachio
Butternut	Black walnut
Cashew	English walnut
Chinese chestnut	Carpathian walnut
Filbert	Pignut hickory
Ginkgo	Shellbark hickory
Hazelnut	

The Farm Orchard

If the farmstead has an orchard, you're probably already eating the fruit in your mind's eye! If I may be so impertinent, I'd like to suggest you turn your mind, and your eyes, toward the more practical pastime of evaluating the trees to find out if the orchard is a valuable asset or a liability.

First, is it a home orchard or a commercial orchard? You won't need to ask because most experts say that it takes about 40 acres of fruit or nut trees to successfully support a family. If you find an orchard that is 10 acres or more, you can plan on making it produce part of your farm income. A home orchard can be from four or five trees to a couple of planted acres.

Next, consider the condition and age of the trees. Trees that have not been well cared for will need lots of hard work, pruning, spraying, and time to bring them back into any kind of paying production. While it's difficult to generalize, an orchard is usually productive for about fifty years. If the trees have been there for generations, it may not be worth your while, work, and money to try to rejuvenate them.

If the trees are new, make sure the farmer has selected a site that has deep, fertile soil that's well drained to plant them. Most tree crops grow best where they can develop deep root systems. However, there is such a thing as soil that is too deep and too rich. Orchards on rich bottom land often grow more foliage than fruit. The best location for orchards is usually a sunny hillside where there is little danger of wind damage or sudden freezes. An orchard needs a good supply of water in case of dry spells.

If you are fortunate enough to run into a farmstead where the seller has been a latter-day Luther Burbank and has used many varieties of fruit and nut trees in his landscaping plan, you should count this as a valuable asset of the farm. You will be able to count on a fistful of blessings and enjoyment from these provident plant pals.

1. They may bring you added farm income.
2. They will bring color, flowers, and fragrance into an otherwise drab landscape.
3. They will provide relief-giving shade.
4. They will provide delicious snacks and desserts.
5. They will help to make your farm a thing of beauty and a joy—for many, many years!

The Farm Garden

The chance to grow a bigger, more bountiful garden is one of the biggest reasons why many of us city dwellers want to come to the country. Most farmsteads will include a home vegetable garden in their plan. It's important that the garden be located on rich, fertile soil. Vegetables will grow fairly well on most soils, but fertile soils will lessen your work load and increase your harvests. If possible, the garden should be located where there is a deep, sandy loam. A pH between 5.5 and 6.7 is best for growing most vegetables. Add lime if you want to grow asparagus, beets, carrots, or spinach.

As I have previously mentioned, you should have the soil on

the farmstead tested along with the other soils in the fields. Contrary to most notions, the farmstead will have to have highly productive soils if you want to have a nice landscape, an orchard, a farm garden, and an attractive lawn. The USDA estimates that 80 percent of your family's food will be grown on the farmstead and that takes good fertile soil.

See that the vegetable garden is located close enough to the farmhouse so the family members assigned to take care of it will have easy access. Daily visits will make it easy to spot and squelch any problems as they arise.

The part of the country where the farm is located will determine what crops can be grown. Rotation planting will insure a steady supply of vegetables for the table and for canning and freezing. Rotating properly will also increase the fertility of the soil and cut down on space requirements. Your vegetable friends will need from six to eight hours of sunlight during the middle part of the day. They will appreciate it if you put the garden where there is protection from heavy and prevailing winds.

If the garden has to be located on a sloped piece of land, see that the seller has been planting it across the slope. If not, and the garden site has been in use for years, you can expect to find played-out soil. On level ground, check to see that the rows run east and west with the shortest growing vegetables planted farthest south in the garden and the tallest growing at the north. This will insure that the plants are getting enough sunlight. Gardens usually contain the plants a family likes best so don't be overly concerned if you spot vegetables not suited to your tastebuds. After all, you probably will get a big kick out of revamping the garden plan as soon as you can get situated in your new farm home!

If the vegetable garden looks healthy and free of pests, you can consider it a valuable asset, which should cut into the high cost of your family's food budget. If it needs lots of work, don't pay any extra for the chance to do backbreaking labor.

Flowers on Your Farm

A country place without flowers would be hardly worth the gas to get there. As you look over various farmsteads, try to pick out places for growing flowering annuals and perennials. Most farm wives will have planted a few roses, azaleas, camellias, lilacs, lilies, stocks, tulips, or zinnias. When you look over their flower gardens, keep in mind that you're going to be awfully busy on the farm. Your flower garden doesn't have to be enormous and it doesn't have to be one that takes several hours every day to care for. Look for plantings that provide flowers, fragrance, and not a whole lot of labor.

If you love flowers and have had a lifetime's success with growing them in your city or suburban garden, you may be attracted to a farmstead with a greenhouse or the ideal climatic conditions for growing flowers outside. You may be encouraged by the seller to think that you will be able to supplement your farm income with a part-time commercial flower-growing operation. I suggest, if you have no experience with this type of specialized farming, that you grow into it. The key to making money with flowers, or other highly perishable farm goods, is to have experience and know your markets. Go slow, gain experience, and get good advice.

If the farmstead has an attached or freestanding greenhouse, it can offer countless hours of happiness and fulfillment spent in the wonderful hobby of growing and propagating flowers, houseplants, vegetables, and seedlings. A good, working greenhouse can provide your home with a continuous supply of colorful plants and flowers all year round.

Check to make sure it's in a location where there is ample sunlight. The reason most flowering plants don't do well in your home is because they don't get enough sun. This is especially true in winter. The location should have a south or eastern exposure so that your plants get the first warming rays of the morning sun and will remain in full or partial sunlight most of the day. If artificial lighting has been installed, the location might not be so critical. If too much sun is present, see if the owner has installed any type of shading.

Good drainage, a proper foundation, full or partial concrete floors and doors that are large enough to move all equipment in and out easily during annual cleanups are all important greenhouse features. Extras like automatic climate control, winter heating, ventilation fans, or air-conditioning are worth money. They could be very costly to add later.

Perhaps most important is the size of the greenhouse. Will it be large enough to fulfill your needs? Is it too large? If so, it may be too costly to heat and maintain. Remember that it is better to have room for expansion than to have to tear the thing down and build a brand new facility that's large enough.

Once again, don't pay extra for a greenhouse you have no intention of using. Especially if you are moving to the country to do your gardening outside.

A properly maintained greenhouse with heat, light, water, ventilation, and good staging can be valued anywhere from five hundred to several thousand dollars for a home and hobby setup. Setups for commercial plant and flower growing can be evaluated much higher. In the latter case, it's best to have an expert tell you if the facility is a good one and worth the seller's price tag.

For more complete basic information write to the University of Illinois Cooperative Extension Service, Urbana, Illinois, and ask for "Home Greenhouses for Year-Round Gardening Pleasure" (Extension Service Circular #879, out of state, ten cents) and "A Simple Rigid Frame Greenhouse for Home Gardeners" (Extension Service Circular #880, out of state, ten cents).

Pools and Patios

Farm families are in need of rest, relaxation, recuperation, and recreation as much or more than their city cousins. Often, having to leave the farmstead to find these amenities and facilities is too impractical, time-consuming, and expensive. One alternative is for you to find a farm that has family-oriented recreational facilities like a pool and patio built into the farmstead plan. If you find a farm with a pool and patio, you will be wise to determine if these have been well constructed, properly maintained, and designed with an emphasis on health and safety.

In the case of the pool, see if the type of facility—aboveground or excavated, metal, fiberglass, concrete, or plastic—meets with your ideas of the kind you want. Is there an adequate water supply? Is there a gas water heater to allow your family to use the pool longer? Check the filtration and pumping system. Try to ascertain if all underground piping has been laid below the freeze line. Check for cracks, tears, or leaks. And check around the pool for any drainage problems. If the pool has a diving board, make sure it has been placed over a portion of the pool that is of diving depth. Most authorities say that ten feet is the minimum depth for a pool diving area.

Look over the entire patio layout. Is it large enough to accommodate the home-oriented outdoor living and recreational requirements of your entire family? Are there separate areas for adults and children? Areas for conversation, relaxation, and play? Is there a cooking area and an area where the whole family can eat together? Is the patio attractive? Does it complement the home? Is it sunny or shady most of the day? If sunny, is part of it covered by an awning or sun roof? If located in mosquito country, is there a screened-in area for the family to use in the evenings? Is there enough lighting to allow for night use? Are the materials that have been used in construction—wood, stone, concrete—of good quality? Is the flooring resting on a good foundation and is it designed for proper drainage?

Will the pool and patio facilities get enough use to warrant the additional capital expenditure likely to be asked for them by the seller? If there are no such facilities on the farmstead, is there room enough for them in case you should want to install them later?

Outdoor Lighting

It's obvious that outdoor lighting is highly important to the safe and efficient operation of the farmstead. In addition to lighting pools, patios, and other recreational areas, floodlights will be needed in work areas and outside of animal housing when the farmer has to work late. It will be needed as a safety measure along walks and driveways and for the easy identification of sudden visitors to the farm.

Experts suggest that instead of running the main power supply to the farmhouse and tapping off distribution lines from there, it should be brought first to a main distribution pole near the central service court and distributed to all parts of the farmstead from that point. All outdoor lighting plans should allow for possible expansion of the system.

Walls and Fences

If you've read a little Robert Frost, you know that "something there is that doesn't like a wall and wants it down." As you inspect the farm, remember that little bit of poetry and take it to heart. Stone walls take days of mending after winter has had its way with them. Those white board fences you see enclosing pastures and edging country lanes take gallons of paint. Even the spare-looking three-wire fences take miles and miles of costly wire. And probably in many instances they could be replaced with single-wire electric livestock fences.

In recent years, many factors have changed the reasons for, and functions of, fencing. The old stone walls that divided the fields of New England farms have been torn down and carted away to allow proper drainage of surface waters and for modern contour plowing and planting methods. Windbreaks and snow fences have been partially replaced with tree and shrub "shelter belts" which are 30 to 50 feet wide and located at a minimum distance of 50 feet from buildings and farm structures at critical corners of the farmstead.

In landscaping the farm lot, the modern approach is to fence in only the *critical areas.* A combined sense of practicality and creativity has brought about a more varied look to the ways privacy can be guaranteed, trespassing prohibited, useable living space enclosed, and unsightly views screened off. As you tour various farm properties, keep in mind the amount of time, labor, and money you will have to put into the upkeep of walls and fences. Try not to get trapped into buying a farm but paying for a paint factory!

If the farm is a livestock operation, or if you intend to convert it to one, fencing will be an especially important factor in a successful operation. Examine carefully all line fences around the property. If they are not in good repair, or sufficiently strong, they may allow your stock to stray off the farm property. Feed lot, or dry lot fences must be adequate for the type of animals you intend to raise and graze. Chutes and corrals must be strong and well maintained.

The cross fences between pasturage and cropland should be adequate to keep animals where they are supposed to be. Sometimes it pays to remove high-upkeep cross fencing and replace it with less expensive single-strand electric fencing or other types of temporary fencing. Consult your county agent about the feasibility of this kind of project.

In the past year or so, fencing wire has increased more than 25 percent in price. Use good judgment. Every effort should be made to keep fencing adequate while at the same time control costs of upkeep and replacement.

AVERAGE FENCE LIFE	
Steel posts	30 yrs.
Snow fences	8 yrs.
Wood fences	15 yrs.
Woven wire fences	15 yrs.

Walks and Driveways

Look over the various approaches to the farmstead. Are all the walks and driveways open, direct, broad, and easy to maintain? By "open" I mean are they highly visible, both to those using them and to the farmstead inhabitants? Has the owner eliminated unnecessary gates and fences which may be costly to maintain? Are the walks wide enough to let two people approach and leave the house side by side (four feet wide)? In snow country, are they made of concrete, asphalt, or other easy-to-clear materials? Are they well lit at night? Are they slippery when wet?

Are the driveways and service courts wide enough to provide cars and trucks with adequate clearances? What about large farm machinery? Are driveways paved with asphalt or concrete? The Farm Safety Council suggests minimum driveway and field lane clearances of sixteen feet. If drives or private roads entering the farmstead are not paved, find out what it is going to cost you to keep them in good condition.

Farm Access Roads

As you turn off the county road toward the farm you are thinking about buying, you may see a sign that says Dead End—No Thoroughfare or Private Road. If so, be extremely careful about buying this farm without investigating all the easement and maintenance problems that may be connected with access roads to the property.

Does the road continue past several farms? If so, you may find that you and the owners of those farms share the upkeep costs of the road. Worse luck, you may find that only the owner of the property you want to buy is maintaining the road! Or you may find out that the road to this particular farm is owned not by the seller but by the owner of an adjacent piece of property. There are many, many land and property problems that can arise over access roads. In order to protect yourself, you must get a legal right of access, or easement, which allows you, your family, friends, visitors, tradespeople, utilities, etc. the right to travel back and forth from the property to the nearest, most accessible public road. In order to obtain these rights legally, properly, and in perpetuity (as long as you own the property), you should have an attorney *obtain a Warranty of Easement in the sales contract and recorded on the deed.* The Warranty of Easement means that the seller warrants that all access rights are being conveyed to you in the sales transaction. You and your attorney should go over this warranty very carefully to be absolutely certain your new farm will have a legal access road directly, either over existing roads or on some future road not yet built (your choice) to the farmstead. Easements can contain many traps, so draw your access road easement out on a map. Then go out and walk or drive over it to be sure you understand exactly what you are getting.

Another thing you should find out is whether any other parties have easements over your land. You don't want to have your chickens go into hysteria when the motorcycle rock band drives through your farm each July to their annual Acid Rock Festival on the county land behind your place!

I have a motto, "Private roads can be a pain!" On some western ranches, the road from the property line to the farmstead can run as long as several miles. Even if the road runs entirely over your property, you must consider the costs of maintaining this road. I suggest you find out from the seller how much he has had to spend to maintain this private access road over all the years he has owned the property. Remember, the Law of Agency protects you from any misrepresentation of negative factors or features by him or his agent. You may find out, after a little digging, that a few years back, "during the big rains," the road was virtually impassable until he spent several thousand dollars for fill and regrading! If the seller has not owned the property very long, try to discover if any previous owners were ever forced to pay unusually high road maintenance costs *for any reason.* Get a competent local road builder to estimate your upkeep costs over a period of years.

The Lawn

A beautiful half-acre or more of growing, green carpeting unrolling down from the farmhouse gives a country place an elegant charm that is hard to improve upon. All those luscious blades of grass, collectively, seem to form a perfect welcome mat for friends and visitors approaching the farmhouse door! This is an especially important feature if you intend to sell any of your farm products on the farmstead itself.

Just about every homeowner wants a nice lawn—59 percent of all Americans include a pretty lawn as an important feature of an ideal homesite. So why shouldn't the farm buyer be searching for the same type of homesite? The ideal lawn ties the trees, shrubs, flowers, garden, walks, driveways—all the elements of the farmstead landscape—together into a harmonious and attractive whole. You will want to survey the farmstead lot to see if such an ideal can be easily realized.

If the seller shares your lawn-loving sentiments, there may already be a well-maintained lawn on the farm lot. If the nature of his farm operation allowed little time for lawn and landscaping chores, it is more likely that the lawn will require some renovation. Early spring and midautumn are the best times to renovate an old lawn. Unfortunately, these are almost always the two busiest times in a farmer's year. Look for a farmstead lawn area that is large, fairly open, free of rocks, underbrush, and complicated plantings. This will cut down on the time you will have to devote to lawn maintenance and renovation. It should also allow you to use labor-saving equipment such as a tractor-type mower with renovator and aerator attachments. Most of these labor-saving tools and equipment are expensive, so if the seller has them, you may want to ask that they be included in the purchase agreement along with the other farm machinery necessary for you to operate the farm successfully.

Before purchasing the farm, you should take soil samples from the lawn area just as from the fields, orchard, farm garden, etc.

Also, have a pH test made of the lawn area soil. Generally, lands east of the Mississippi will need liming in order to release the fertilizers and plant nutrients held in the soil so they can be absorbed by the root system of the grass.

Sometimes farmers who have always taken great care to select proper seed for cropland planting will not do this when it comes to putting in grass. Seed that is not suited to the climate and growing conditions of the farm location may sprout, but it will cause you all sorts of maintenance headaches. It is important that you select a farmstead lawn that is as easy to maintain as possible and that you include a regular program of lawn maintenance in your farm calendar.

A Landscape Balance Sheet

Now that you've examined all the features of the farmstead landscape, you will need to go over any mental and written notes you have taken in order to make a balance sheet of the landscaping assets and liabilities. Coupled with the farmstead map you have drawn, this balance sheet will help you decide whether or not you still want to buy the property.

The farmstead map and landscape balance sheet will also help you formulate a long-range landscaping plan, if and when you do buy the farm. This plan should incorporate a regular maintenance program and a schedule for any future changes and improvements you may want to make. These should be scheduled in order of priority. By laying out your work schedule and extraordinary cash expenditures in this planned way, and following the plan in a persistent manner, you will be able to create an attractive and satisfying home landscape in the most efficient way.

Fill Your Ark Carefully

The farmer who owns the country place you are considering will more than likely have a lot of livestock he'd like to unload along with the land and the buildings. If he had the place for several years, you will probably hear more "*moo-moos* here, and *quack-quacks* there" than on Old MacDonald's farm! So while it may be tempting to take the place as is, with all the livestock thrown in, I suggest you go slowly. Take a tip from Noah and fill your ark carefully, or you may not be able to stay afloat very long.

When purchasing the stock for your new farm, you should have a plan. The plan will have to be tailored to your specific situation. The type and amount of livestock you purchase should depend, primarily, on the size of the farm and whether you intend to be a part- or full-time farmer. Generally, the more productive land you have, the more animals it will be able to support with feed and pasture. The less time you can spend on farming, the fewer animals you will be able to supervise. Of course there are other factors which must be considered. The more operating capital you have for your first year, the better your chances of running a large-scale livestock operation successfully. Also, the better your management methods and techniques, the better your chances of maximizing the total number of animals on a small farm. Specialized operations, like shed dairies, can allow you to keep as many as two hundred head of cattle on less than ten acres.

Obviously, it is very difficult to generalize about the size of livestock operations. Each farmer tailors his herd and hutches to his farm goals and income needs. However, a beginner should follow the basic guidelines of common sense. Keep in mind that some animals work for you; *you* will have to work for others. In the beginning I suggest you purchase the animals that will add to your farm income in the shortest possible time. If you sometimes think your kids are going to eat you out of house and home, just watch how quickly nonincome-producing animals can gobble up your cash reserves and eat you right off the farm!

If you have the experience and are going into full-time farming, you will probably purchase a farm that is a going operation. Perhaps a dairy farm with thirty to forty freshened cows, milking equipment, grain storage, feed processing facilities, and plenty of pasture. Or maybe a going egg farm with 2,500 to 3,000 laying hens, off-the-floor caged housing, artificial lighting, and an efficient feeding system. Both of these operations share one important thing in common; the animals you purchase will be producing farm income from the very first day you move in and take over! In each case, however, the smart farm buyer would carefully check the production records of the cows or hens to be certain they are producing up to a certain acceptable standard. Buying quality stock that has been properly managed is the key to your chances for success. According to the Bureau of Agricultural Economics, annual milk production per cow should range between 8,000 to 12,000 pounds. A good laying hen should produce about 250 eggs per year. Poor producers in both dairying and poultry raising should never be purchased. If they are purchased, they will eventually have to be culled from your stock if you are to have a successful operation.

While dairying and poultry-egg production can give you a farm income that's distributed evenly over the year, other types of livestock will only produce income after a fairly long period of management. With reasonably good management, hogs or sheep will produce income within six months. Hogs are particularly adaptable to diversified farming operations and they are efficient in turning feed into meat for a minimal cost to the farm operator.

Horses

Anyone who has ever been to the racetrack knows horses can be a hardship! This is equally true for the new farmer. My friend John Means, who has been raising and training saddle horses for most of his adult life, says the new farm family members who want horses should be aware that they require a high investment in money for the original purchase, feed, and maintenance. They will also require considerable investment of your time to train, exercise, and groom them. John, who happens to love horses, says that unless you happen to have land close in to the city where you can operate a riding stable, horses are not likely to bring you any immediate farm income.

Raising breeding stock or racehorses are enterprises which require great cash reserves to handle the high operating costs and to compensate for losses due to injury or illness. It's no wonder horse racing and breeding is called "the sport of kings."

When starting any kind of livestock operation, remember that each time you increase the number of animals you keep, you will also increase your operating costs and the risks of disease and infection which could destroy all your stock. Go slowly. Grow into a big, large-scale animal operation.

AVERAGE LIFE SPAN FOR FARM ANIMALS

Dairy cows	8 yrs.
Goats (breeding)	5 yrs.
Hogs (breeding)	5 yrs.
Sheep (breeding)	5 yrs.
Horses	10 yrs.

If you are not interested in farming full time, the number of farm animals you keep will be dependent to some extent on the amount of land you purchase. It has been estimated that a fertile 10-acre farm will be able to support a cow, twenty-four hens, and a pig. Any more livestock will rely on outside sources of feed supplies. Let's consider some of the more typical classes of livestock that may fit into your subsistence-type farming enterprise.

Cows

Keeping a cow or two for your family's milk consumption can represent a considerable investment, but it may be justifiable in terms of improved nutrition and some savings on the family food budget. For a family of four, it's probably best to stick to one of the smaller breeds like a jersey or guernsey. These two breeds also produce milk that is highest in butterfat content, which means it is a high-grade product.

You will have to arrange to have your cow bred every year. The service fee for this can range from nothing to twenty dollars, depending on who you know and what kind of calf you want. In terms of actual cash, the calf won't be worth much more than the service fee and the cost of its feed. If you only keep one cow, there will be about two months of the year in which she is not lactating and will not produce any milk. For that reason, it may pay to keep two cows and have them bred at different times of the year.

In order for your cow to provide an economical family milk supply, your farm property should have about two acres of fairly good pasture. In addition, you will probably have to produce or purchase a couple of tons of hay and about 2,500 pounds of grain and feed concentrates.

When selecting new stock, the pedigree, production record, and individual physical appearance should be the three most important factors. Your dairy cow should be in good general health and have a good disposition. Always deal with reputable farmers or dealers when buying any type of farm animals.

Please don't make the mistake of assuming you can successfully operate a dairy on a part-time farm. Twenty to fifty cows will take up a good eight hours of your time, every day, even if you have the most modern facilities.

Beef Cattle

My friend Jerry Noland and his wife, Marian, always keep a few head of registered Angus purebreds on their country acres down in Ocala, Florida. While they have sold some beef, they are primarily interested in producing their own family beef supply. When Jerry fixes you a sumptuous Sunday brunch of steak and eggs, you can be sure those are Black Angus steaks, pardner! But Jerry admits that there's no point in kidding about it; when the year is over, it always ends up costing him a great deal more to raise his own beef on the hoof than if he had bought it already butchered at the store. On the other hand, Jerry claims, raising purebred cattle is a pleasurable pastime which offers great returns in terms of pride and satisfaction.

If you have your mind set on raising your own beef supply, there is probably nothing I can say that will get you to change it. Go ahead and keep a cow and calf. They will need about 3 acres of good pasture to get them through the summer. If more pasture is available, remember the tried-and-true cattleman's rule and *don't overgraze*.

John Polich, a well-known California TV director who does the Kings' and Lakers' games, says the beef he raises for family consumption needs about 5 acres of pasture per head. (Of course, the amount of pasture you will need per head will vary according to the part of the country you farm in.) John says his pasture tends to dry up during the hottest summer months and then the cattle must be supplied with supplementary feed from off the farm.

During the winter, your beef cattle will need hay and some grain supplements. Experts suggest about one to one-and-a-half pounds of protein supplement per head per day. Both John and Jerry say you should give your beef cows plenty of water and a supply of salt and bone meal for minerals. I suggest you consult your state ag-extension service for recommended feeding programs in your area. The extension service and the various cattle breeders' associations will be happy to supply you with information regarding the selection, purchase, and management techniques.

Among the many excellent books that the potential cattleman should read are *Beef Cattle*, by Roscoe Snapp (John Wiley & Sons, New York, 1969); *Beef*, by Harrell DeGraff (University of Oklahoma Press, Norman, Oklahoma; *Cowboy Economics*, by Gen. H. L. Oppenheimer (Interstate Printers & Publishers, Inc., Danville, Illinois, 1971).

Jim Harper, a well-known investment counselor and business

manager for show-business celebrities, has been in and out of various phases of the beef cattle business all his life. He says that right now many of the small-scale beef-feeder operations are having trouble beating the price cost squeeze and staying in business. He advises you to stay out of the business unless you become partners with someone who has plenty of experience, plenty of operating capital, a thorough understanding of cost-return analysis, and marketing techniques. After listening to Jim, I wonder what a guy like that needs with you and me? Jim also advises that you thoroughly investigate the current developments in the tax laws before investing in beef cattle breeding. If the government classifies your operation as hobby-farming instead of as a business, your tax savings will be nil.

Hogs

As I've said before, hogs are very efficient at turning feed into meat. In order for your swine operation to be successful, you will need good quality stock from a reliable breeder. You will also need to have a dependable supply of good quality, low-cost feed and a cost-conscious but adequate feeding program. Very few part-time farmers keep a sow. Instead, they are more likely to keep one or two feeder pigs which they buy in the spring, feed up to 250 pounds, and have butchered in the fall. The key thing to remember is that nearly 80 percent of your operating costs will be feed. Corn, grain sorghums, wheat, barley, skim milk, garbage are just a few of the food sources you should consider. Bone meal, tankage, or fishmeal will supply necessary minerals. It takes about 1,000 pounds of feed to produce one finished hog ready for the butcher. By acquiring a good knowledge of various feed sources and the nutrient content of various foods, you will be able to lower those requirements considerably and your costs along with them. It may be that you can make a deal with the local school or hotel to buy their garbage.

My friendly insurance man, Charlie Rose, remembers when

he was a boy living on a dairy farm in the Ozarks. He and his dad used to take the Model-T truck across the state line to a Kansas distillery where they could purchase the used mash for ten cents a barrel. Charlie says you want to talk about "hog-wild"—those pigs went crazy for old rotgut sour mash!

If you watch the hog market reports, you will be able to buy your pigs for about twenty-five dollars, sometimes even less. If you have a friendly neighbor, you might even be able to persuade him to give you a pig or two for nothing.

If you intend to go into swine production to make farm income, you will have to pay more careful attention to your hog selection. Selection of meat-type breeds like Hampshires, Poland-Chinas, Durocs, Yorkshires, San Pierres, Landraces, Tamworths, and Berkshires will almost always bring higher market prices. In recent years hogs of various crossbreds have become popular with producers. I suggest you become as familiar with the various breeds as you can, consult with local producers, and with your county agent. Your state extension service will have numerous bulletins on swine production to aid you. *Approved Practices in Swine Production* by J. K. Baker and E. M. Juergenson (Interstate Printers & Publishers, Inc., Danville, Illinois, 1971) is an excellent book for beginning hog farmers.

Sheep and Goats

The largest bands of sheep are raised on huge western ranches. Any more than that, I don't know. For information about successful sheep ranching techniques write to the USDA or your state ag-extension service for bulletins. Check with your county agent for advice. Excellent books on the subject are widely available. A few of these, which will be helpful to all types of sheep producers, are *Sheep Production,* by Clarence E. Bundy and Diggins (Prentice-Hall, Inc., Englewood Cliffs, N.J., 1958); *Sheep Science,* by Kammlade (J. B. Lippincott, Philadelphia, Pa., 1947); *Livestock & Poultry Production,* by Clarence E. Bundy and Diggins (Prentice-Hall, Englewood Cliffs, N.J., 1974).

If your family likes lamb stew, lamb chops, or leg of lamb, there is probably some justification for keeping a few sheep for home meat production. I suggest you ask the manager of the local feed store to recommend a breeder or farmer who is reputable and who will sell you a couple of two- or three-year-old ewes or some feeder lambs. The best time to buy feeder lambs is in the late summer or early fall. You may want to buy older ewes, if the price is right. Before doing this have them examined by an experienced sheepman. Stockmen use their hands to go over the animal carefully. Generally, the basis for selection is the animal's overall body conformation. Other factors in selecting good sheep are: wool type, breed, health, age, soundness of mouth and udder, quality of flesh, and uniformity of size and type. Sheep-breeder associations will be happy to furnish you with a scorecard for judging rams and ewes of their breed. Regarding wool type, it is usually agreed that medium wool type sheep make excellent meat producers.

Sheep are excellent grazers. If there is a prevailing wind away from your house, you may even use them to keep your lawn trimmed! It's important, however, that you rotate pastures to keep your sheep from overgrazing. The same feedman who told you where to buy your sheep will recommend a good supplemental and winter feeding program. Sheep alone are probably not a good risk for small-scale or part-time farmers. You might want to combine them with fruit or goats.

Goats can be pastured with your sheep. People who raise them tell me they are the perfect dairy animal and that their milk is unsurpassed in taste and quality. As a teen-ager, I had a near-disastrous experience with goats. A man named Jones who had a local goat farm asked me if I'd be interested in baby-sitting his goats while he went on vacation. Knowing nothing about the animals, I readily agreed. One evening I turned them out to pasture—in the wrong pasture. Around one or two in the morning a state trooper knocked on the farmhouse door to tell me that my goats had got out and were spread out all over the interstate highway at the foot of the hill! It only took me a few minutes to get dressed and down there, but it was hours before I had all those vagabonds back in the right field. This is my voice of experience telling you that goats in pasture require good fencing.

Experts say that you should turn your goats out to pasture gradually. Start by giving them a full ration of hay before turning them out. Then gradually reduce the hay ration until they are feeding almost exclusively on pasturage. Goats are good browsers and will eat underbrush. If you are not careful, they will also eat the bark off your young saplings. When not on pasture, a doe needs about two pounds of leguminous hay and about two pounds of mixed grain. Goat does go dry about a month or two before kidding. During this period you will probably have to feed them supplementally. Legume hay is the main ration with a commercial feed mixture of corn, bran, soybean meal, molasses, salt, and minerals. If you want to mix your own home-grown feed and avoid feed store prices, I suggest you write to your state extension service and ask for recommended mixture recipes.

Since goats are not universally popular, you may have trouble finding quality animals in your area. I suggest you write to the American Dairy Goat Association (c/o Kent Leach, Box 1908, Scottsdale, Arizona 85252). They will send you information plus the *Dairy Goat Journal* which contains information for buyers.

During that teen-age goat-sitting adventure that I mentioned above, I also had some experience with milking goats. This too was nearly catastrophic! Goats have been called "the poor man's cow," and not without reason. The only trouble was that I hadn't had too much experience, and it seemed like I got more milk *on me* than in the milk pail. Finally, it occurred to me to move the milking stand into the cow barn and use the automatic milking machine. I've learned since that time that cow-milking machines have been adapted with attachments for goats, but that wasn't the case then. When I hooked the first doe up to the machine it nearly tore her udders off! Be careful about letting unsupervised teen-agers become your goatherds. Goats with experienced, gentle milkers—like you will soon be—will give 5 to 7 pounds of milk a day. A good milking doe will produce an average of 2 quarts a day for eight or nine months of the year. By staggering the breeding of your does over a period of several months, you will be guaranteed a steady supply of milk.

If you only are going to have a small place in the country, I'd advise against your getting a buck. Remember the old saying, "Strong like bull, smell like goat!"? Well, believe me, it's talking about *buck goats!*

Poultry and Eggs

The farmer who owns the place now, no doubt already raises some chickens. If you have inspected his housing facilities and found them to your liking, you may want to look at his flock to see if the chickens are the kind you want to raise.

Intensive breeding and research have been responsible for developing several breeds of chickens for each phase of poultry farming. The most efficient breed for egg laying is generally acknowledged to be the White Leghorn. A majority of all the commercial laying flocks in the United States are White Leghorns or crossbreds. If you are inexperienced in the poultry business or if the farm is located near a noisy airport or highway, I'd advise against purchasing White Leghorns. They have a reputation of being nervous-breakdown-prone and must be expertly managed or production will fall off drastically. There are other layers just as satisfactory for the small-scale commercial operator or subsistence farmer. Among them are: Rhode Island Reds, White Plymouth Rocks, and various hybrid and crossbred strains such as the California Gray–White Leghorn crossbred or the Rhode Island Red–White Leghorn crossbred.

Other popular American breeds like the Wyandotte, Barred Plymouth Rock, New Hampshire, and New Jersey Black Giant are used in pure strains or crosses for both egg and meat production. The English White Cornish and its various crosses are the most popular breeds for broiler production.

If you have a traditional streak, you may take a tip from my friend Patsy Nelson and keep a few old-fashioned Bantams, or "banty hens," as Patsy calls them. Victor Sen Yung, my Chinese food advisor, says that young cockerels of the Cochin breed make the best spicy Szechwan chicken with tangerine peel you've ever tasted! Maybe you'll want to give your food a British flavor with Sussex or Red Caps or raise Minorcas or Anaconas for your eggs. If you raise your chickens by the batch, you will have to replace all except well-culled hens in a year. Buy your day-old chicks from a reliable hatchery or through a reputable local feed store which deals with breeders.

It is possible to buy started pullets, but nothing is more fun for your family than starting your flock from day-old chicks! If you locate out in "the country," get straight-run or unsexed chicks. If you live closer to town and are more interested in eggs than meat, you can order sexed pullets. The cockerels have a way of learning to crow and irritating your neighbors!

If you want a flock of two dozen layers, get a hundred chicks. That way you can pick out the best-looking pullets for layers. Get debeaked chicks if you are worried about cannibalism. This shouldn't be a problem if you provide adequate space and maintenance.

If you are planning to raise chickens for meat, be sure and tell this to the hatcheryman where you order your chicks. He will probably give you White Cornish–White Rock crossbreds or another of the meat-producing hybrids. Meat birds are generally classed according to size and age. *Fryers* are birds which weigh less than three pounds and are about eight to ten weeks old. *Broilers* are about the same age, but are larger, weighing three to four pounds. *Roasters* are birds weighing more than four pounds. They are often much older than ten weeks. Cockerels weighing more than five pounds are mature and will soon discover the art of crowing. They should be slaughtered before or as soon as this happens.

Chickens need a dependable supply of fresh water every day.

Start day-old chicks with a one gallon chick fountain per hundred chicks. A small flock of two dozen will use about 250 gallons of water a year. Your chances for success with poultry depend on purchasing good quality stock and then setting up and following a proper feeding, management, and sanitation program.

Your farm agent, feed store, state extension service, hatchery, USDA bulletins, and quite a number of books can help you avoid numerous pitfalls due to mismanagement or disease. Perhaps your greatest help will come from the advice and experience of the local poultry producers in the area. One helpful booklet is the *Purina Commercial Poultry Program,* which can be obtained from the Ralston Purina Company, Checkerboard Square, St. Louis, Missouri 63188. Start small and grow into commercial production. A few years' experience in handling a small flock will stand you in good stead as you expand. Chickens are susceptible to a great number of diseases which can be fatal to your flock. Each time you expand your operation, you take the chance of introducing disease. Proper sanitation procedures will go a long way in keeping your flock healthy and productive.

Poultry production is one farm enterprise that is very well suited to the part-time farmer. You should have no trouble maintaining a flock and keeping a full- or part-time job if you lay out your program and the connected chores properly. Chicken farming can be followed by folks from eight to eighty—if they have a good sense of responsibility. Get your family involved. Be specific about who is responsible for what. Chickens are negatively affected in terms of egg production if their routine is disturbed in almost any way.

Turkeys, Ducks, and Geese

You will find that advice and information is readily available if the seller should offer you his stock of turkeys, ducks, or geese. All three birds represent a growing commercial market, or you can raise them along with your other farming pursuits.

At one time, the only real market for turkeys occurred around the holidays. Now turkeys have become a year-round meat item stocked in most supermarkets. In 1974, well over 100 million birds were produced commercially. However, turkey growing is a highly specialized business and it will be necessary for you to know what you are doing and secure your markets before going into it full time.

Turkeys are very sensitive to disease and disturbances. They don't like to be moved, so if you are not buying your starter stock along with the farm, try to select good quality poults from a turkey farmer near where you intend to locate. In that way you won't have to move them very far.

For selecting turkeys, the most common varieties are: the broad-chested Bronze or White, the White Holland, the Narragansett, the Black, the Bourbon Red, and, in recent years, the Beltsville Small White.

Raising one or two turkeys on the farmstead for your own holiday meals is not normally recommended, as the birds do not do well on the same farms where chickens or other types of fowl and poultry are kept. If you have intentions of starting a turkey ranch, visit as many successful producers as possible. Thoroughly analyze their operations and seek their advice. Don't expect to get immediate income from your birds; it takes at least six months to finish a prime turkey broiler.

Write your state ag-extension service for information about getting started in your area. Contact the USDA and the following state's extension directors for information in circulars and bulletins: University of Minnesota, St. Paul 55101, University of California, 2200 University Ave., Berkeley 94720, Virginia Polytechnic Institute, Blacksburg 24061.

Ducks and geese are much easier to raise on a small scale than turkeys. And a roast duck or goose will give your holiday meals a different taste, helping you break out of the turkey mold.

You gardeners will appreciate the weeding abilities of geese and the fact that both ducks and geese love to eat snails and slugs! If there aren't any of these fowl critters on the farm now, you will want to get at least a few for meat and eggs. One of our neighbors, Bonnie Johnson, uses hollowed-out goose eggs to create incredibly beautiful handcrafted decorations.

Buy your ducks or geese from a reliable breeder. The most common varieties of barnyard ducks are: Pekin, Muscovy, Call, Cayuga, Black East India, and the ever-popular white Aylesbury. The Emperor and Toulouse are the farmers' most favorite breeds of geese. It's a good idea to start ducks or geese as matched pairs, as these animals mate for life and it will be difficult to bring in a lover later.

Shooting Preserves and Game Birds

When he heard I was working on this book, my friend, actor Ed Nelson, told me to be sure to mention the need for commercial shooting preserves and game bird breeding farms. Ed says there are millions of men and women who like to take to the fields on a crisp autumn morning with a shotgun and an empty pheasant pouch. The trouble is, it's become pretty hard to find a decent place to hunt these days. And when you do, the chances are slim that you will find enough native wild birds to fill that pouch. If you should decide to go in for this type of specialty farming, you'll find you can have the satisfaction of living in natural surroundings and increasing and improving American wildlife. Ed Nelson says, "Controlled hunting is good hunting. It can be just as exciting as that found on public lands." With the tremendous increase in leisure time, it seems certain the need for preserves and for the birds to stock them will continue to grow.

If you're a hunter who is thinking about starting your own game farm, you will be wise to hunt on as many of the hunting preserves in your state, or nearby states, as possible. Try to find out as much as possible about the operations where you hunt. Try to determine the good and bad aspects of each operation and what, if anything, you would do to improve on it. Hunting and fishing magazines often have lists of such preserves.

Contact the U.S. Fish and Wildlife Service, Department of

the Interior, for information bulletins. Probably more important, contact your state department of wildlife and conservation to find out the rules of the road and if there are any restrictions against your starting a shooting preserve. Make sure the farm you pick out to buy is located in an attractive hunting area and is a place which weekend hunters from large metropolitan areas will find easily accessible. It has been estimated that there are nearly 20 million game birds which have been bred for shooting preserves in the United States. Breeders feel justifiably proud that they have done a great deal to encourage controlled hunting and improve the species of quail, chukars, pheasants, mallard ducks, and ornamentals.

Game birds can be raised for shooting purposes or for meat. To find out about the costs to set up a preserve and to operate a breeding farm, contact the North American Game Breeders and Shooting Preserve Association, Inc., Goose Lake, Iowa 52750. For one dollar and mailing costs, they will send you *Commercial Game Bird Management*.

If you write to the Ralston Purina Company, Checkerboard Square, St. Louis, Missouri 63188, they will be happy to send you information on birds and a very helpful booklet entitled *The Purina Game Bird Book*, which gives information on breeder management; growing game birds; game bird feeding programs; starting a commercial shooting preserve; game bird diseases; and additional sources of valuable information.

Honey from the Forrests

Steve Forrest, the motion picture and television star, claims that the most underrated of all the animals that man has domesticated is the honeybee. As more and more woodlands are being cut down and put to other uses, the pollinating insects that live in them are killed or greatly reduced in numbers. Because of this, farmers who raise vegetables, flowers, and fruit have become increasingly dependent on the tremendous pollinating activities of the honeybee.

Steve is superenthusiastic about beekeeping and says that anyone with a little spare time and a little patience can easily keep a couple of colonies on a small farm. Information on beekeeping is easy to come by. Check your local library for *Starting Right with Bees* by A. I. Root. Or contact the Bee Culture Branch of the National Agricultural Library, U.S. Department of Agriculture Beltsville, Maryland 20705. There are bee suppliers in several states. Out here in the Midwest, where I live, Dadant & Sons, Box H, Hamilton, Illinois; the Walter T. Kelly Company, Clarkson, Kentucky 52726, and the A. I. Root Company, Medina, Ohio 44256, are three of the best-known suppliers of bees and equipment for the beekeeper. If you start small and learn as you go, it won't be long before you'll be able to produce enough honey to supply your family and friends. Or, like Steve and his kids, maybe you'll be able to sell some.

It is possible to collect your own bees. But it's not as easy as you might think to follow a bee home to his hive. Bees have some tricky maneuvers to discourage bears and other hive predators, including humans. Probably the best way to collect working bees is to have your friends and neighbors scout for them on their property. Most people are afraid of bees, so they will be delighted to have you come and take a wild hive. They don't realize how hard those bees are working for them in the garden landscape.

Steve says the first time he collected a wild hive a friend called from several miles away. Since he didn't have a beekeeper's hat with a net, he borrowed a wide-brimmed netted hat his wife happened to have from a few years back. He took his gloves and smoke gun and drove to where the honeybee hive was located. Steve says he had no trouble calming the bees with the smoke, getting them into a box and into his station wagon. But the trouble began on the way home with his "smoked" cargo. He was just driving into Hollywood when he heard a buzzing from the back of the station wagon. Pretty soon a bee was flying around his face, looking for a place to land. Then more of them flew up to the front seat to see where he was taking them! Steve decided not to pull over in traffic and start the smoker. Instead, he lit up a cigarette and reached over and put on his wife's hat and pulled down the netting. And so, puffing furiously on the cigarette, he continued on home. Steve says the hat and smoke made him feel safe and protected from the bees—but he wasn't so sure about some of those strange-looking guys who smiled and winked at him every time he stopped for a red light in Hollywood!

Rabbits and Nightcrawlers

George La Fountaine, the best-selling novelist, is an avid fisherman and gardener like me. One day we were fishing from George's beautiful Grand Banks cruiser off the Santa Barbara Channel Islands. The sea bass and yellowtail were jumping all around the boat, but unfortunately we had run out of anchovies and could only stare wistfully at the broiling waters and talk about better days.

Somehow we got on the subject of nightcrawlers and red worms as the ideal bait for freshwater fish. I told George that the best place to grow fishworms is in his garden compost heap. George said he'd like to try it so I promised to bring him a starter batch of Michigan worms the next time I came to California. You see, I believe that Michigan worms catch the most fish. Anyhow I tried. I even bought some of those special boxes they use for shipping worms. The trouble was, every time I got to the Golden State, I would leave George's box of worms somewhere. Seems like I left those worm packages everywhere I went. Now it didn't seem practical to go back and ask waitresses, stewardesses, cab drivers, or program chairwomen for the local women's clubs if they had happened to see a box of worms, so I'd apologize to George and promise to do better next time.

In the meantime, someone told George that the best compost for raising fishworms contained rabbit droppings. So, he went ahead and purchased a buck and a doe of the Netherland Dwarf breed and constructed an elaborate hutch on stilts with a half-inch hardwear cloth flooring over his compost pile.

About a year and a day later, George still didn't have any Michigan worms, but he did have lots of bunnies—fourteen to be exact! He said he would have had four more but the doe scattered her first litter, killing four of the six. He was beginning to wonder whether rabbits made good fish bait when I suggested that he raise some bunnies for fryers.

George told me that when he was a small boy the little Italian lady who lived next door to his family raised rabbits for cooking. He would go over to her house every day after school to feed and play with one with whom he was particularly taken. He says he still can vividly recall the nice old lady smiling at him and saying one day, "You lika dat rabbit?" When he allowed as how he did, she said, "Then, I'ma gonna givit you!" Thereupon, she reached into the cage, grabbed the rabbit by the ears, swung it violently around in the air wringing its neck, and slammed it against a tree. Smiling, she handed him the remains! George says he knows she meant well, but ever since then he hasn't had much of a taste for fried rabbit or rabbit stew! It looks like he's really got himself a problem; he likes rabbits but they don't make him hungry. Funny, if I remember correctly, he likes fishing but doesn't eat what he catches either. I guess the moral of all this is that you shouldn't buy rabbits when you want to raise worms!

If you do want to raise rabbits, either for food or to sell, I suggest you find a reliable breeder with healthy, quality stock. There are about fifty breeds now being raised commercially. You may find it wise to join one of the many breeding associations or clubs. Members often exchange helpful information about management and marketing techniques. Go slowly until you know what you are doing.

The Ralston Purina scientists say that the selection of prime animals is the key to success in the rabbit business. When you purchase your starting stock, they advise that you check all management records which give information on heredity factors, size of families produced, size of animals produced, and the growth potential of individuals in each litter. If the breeder has not kept such records, they suggest you find another who does. *The Purina Rabbit Book,* which may be obtained free by writing to the Ralston Purina Company, Checkerboard Square, St. Louis, Missouri 63188, says it is a mistake to depend too much on appearance when selecting stock. Instead, they suggest you choose your breeding stock by using the following criteria:

1. Litter size
2. Regular breeding
3. Milking ability
4. Litters should average 3 to 4 pounds at four weeks and weigh 4 pounds per fryer at eight weeks
5. Litters which convert feed to meat efficiently
6. Litters with high dress-out percentage
7. Good coat quality.

California is one of the world leaders in rabbit production. You can gain valuable information about how to start a small rabbit ranch by writing to the California Ag-Extension Service, University of California at Davis. The USDA also has several bulletins on housing, health, and maintenance. If you are just raising a rabbit or two as pets, I recommend you read *Enjoy Your Rabbit,* which can be obtained through The Pet Library Ltd., 50 Cooper Square, New York, N.Y., 10003. Do your homework and you'll avoid most of the usual pitfalls.

Chinchillas

Jim Harper, whom I mentioned earlier, has given me a lot of interesting information for those of you who intend to try raising chinchillas. Jim and Rusty Maynes are partners in the largest commercial chinchilla ranch in the world, at Anza, California. The Anza Royal Chinchilla Ranch is located on about five acres at 4,200 feet altitude. The animals need a dry climate in order to thrive. Jim says that by introducing sound management techniques they have been able to keep the yearly food and maintenance costs down to about twenty-five dollars per head per breeding female. This was accomplished by building up the operation to the proper economic size, a process that has taken about six and a half years. He says a beginning producer shouldn't expect to make big money very quickly. He and his partner believe that in order to maximize profits, you need approximately six thousand head. They currently have about seven thousand of those soft-furred little rascals!

According to Jim and his partner Rusty, chinchilla ranching is not complicated, but it requires constant culling of the stock in order to maintain the quality of the herd. Their operation eliminates about 20 percent of the breeding herd each year through a combination of culling and death loss. Even small-scale producers will have to cull this high a percentage of their animals if they expect to be able to produce high quality breeding animals and prime pelts. A neophyte chinchilla producer should buy his start-up breeding herd from a very reliable rancher as it is very easy to get *bad* or even culled animals.

You can start your chinchilla farm in a garage or shed which is not in use. The animals will need cool, dry weather in order for their pelts to prime out, so you may want to investigate air-conditioning for your housing. Commercially sold cages, or holes, come in units of three, four, or six. They commonly sell for about fifteen dollars per hole. You will need two or three cages for each female's offspring.

A young female chinchilla can cost you anywhere from $20 to $250—you get what you pay for. It will cost about $75 to set one female up into breeding. Growers normally breed one male chinchilla to every six to twelve females. The male has a tunnel which leads from his cage to the cages of the various females. Talk about relegating the female of a species to nothing but sex objects! The female chinchilla will have about three babies a year. These animals have a gestation period of 111 days, about four times longer than rabbits. In case you're wondering, chinchillas are not a type of rabbit, as I once believed until Jim set me straight. There are "chinchilla rabbits," but they are merely named after chinchillas because they have soft fur similar to chinchillas. The chinchilla is a rodent which is related to the guinea pig and the porcupine, resembles a small rabbit and has a squirrellike tail. It grows to about ten inches long. The ears are fur-less. It is the fur which is their most remarkable feature. It grows to about an inch and a half long under proper management conditions and is a striking bluish gray with dusky markings. It has been called the most feminine and most luxurious fur in the world. Jim says if you want to turn a woman on, just stroke her face with chinchilla fur. Now that certainly would not be classified as giving her a cheap thrill—chinchilla coats commonly sell for as much as $15,000 to $20,000!

Chinchillas are native to the Peruvian Andes, where they were trapped and made into robes by the Indians. The story goes that one of these robes was "appropriated" by a conquistador captain, who sent it back as a gift to Queen Isabella. Her Majesty is reportedly the first, but certainly not the last, high-toned lady of birth to flip out for her chinchilla wrap! This one casual gift to a queen was the beginning of an exclusive fur trade. Chinchilla became "the fur of royalty." Then, near the end of the last century, this trade suddenly swelled to enormous proportions. The result was that chinchillas nearly joined the long list of animals made extinct by man's thoughtless exploitation. It wasn't until sometime in the 1920s that the animals began to make their comeback on government-sponsored breeding farms in Peru and Bolivia. Commercial growing came to this country thanks to an American engineer who loved animals and brought eleven chinchillas to California.

Jim Harper says that since chinchillas live in the high mountains in nature, sometimes at elevations of 7,000 to 20,000 feet, they like cool, or cold, dry climates. Commercial growers usually raise them at temperatures no higher than the mid-70s. Otherwise their valuable fur will not prime out. The Anza-Royal chinchillas on Rusty's ranch are maintained in an artificial climate. It takes from one to four months to prime out an animal. For this they are placed in a "cold room" where the temperature is gradually reduced from 72 degrees to 42 degrees over a two-week period. It is then maintained at 42 degrees for about two weeks. After that, the temperature is gradually increased over a three-week period until it reaches 58 degrees. The animals are then placed in a "finishing room" at that temperature. Jim says it takes from two to four months for the animals to be finished and the fur to reach the proper sheen, density, and loft characterized as "prime."

It will cost you about a penny per head per day to feed your herd. You should be able to find low-cost commercial feed mixes which will prove economical. Good nutrition, medicated water, and sanitary conditions are essential to a sound program of preventative medicine and maintaining a healthy herd.

As with any specialty type of farming, you should investigate and affiliate your operation with one of the existing marketing services *before* going into the chinchilla business full time. Learn all you can from respected producers and breeders. Visit their operations to see how they have solved the same problems you will have to deal with on a smaller scale. Experience is the best teacher, but you needn't experience a major catastrophe in order to learn what not to do. If you start small, and gain experience as you go, you will soon have a healthy and growing business.

Seed, Feed, and Silage—What to Buy

When you buy a going farm, there will no doubt be valuable crops growing in the fields, valuable grain in storage, and feed and silage for feeding the farm animals. It may be difficult for you to accurately appraise the value of these farm resources. If you do not feel that you should rely entirely on the seller's asking price as the only evaluation, you should seek expert advice. A friendly neighboring farmer, a qualified seedsman, the county agent, the local farmers' cooperative, and your state extension service are all logical places to seek advice. If none of these sources wants to stick its neck out, ask for the name of an experienced farm appraiser. He will be able to tell you if the asking price is fair.

If you are not planning to farm full time, or if you are not planning to keep as large an inventory of livestock as the seller, you should be cautious about buying seed grain or feed which you will not be needing. The seller may tell you that the seed grain will be easy for you to market. He may point out that prices will be better after harvest or, if the grain is already in storage, at a later date. All his representations may be truthful, but you should still be cautious. Marketing seed grain for feed, just like marketing all types of farm produce, requires certain skills and know-how that come with experience. There is no point in your going out of your way to get burned as you learn.

The sources I have listed above will be able to give you your seed and feed requirements for the type and size of farm operation you plan to manage. If you send a sample of the seed grain to your state experiment station seed-testing lab, or to a commercial lab, they will test it for a nominal fee. It's usually best if you do this several months before planting time, so they can expedite your test.

The seed test will tell you about the purity of your seed sample, its ability to germinate, its weed seed content, and the percentage of other crop seeds it contains. If you multiply the percentage of pure seed by the percentage of germinating ability and the number of pounds of seed you are being offered, you will get an idea of the number of pounds of pure live seed that you can expect to germinate. Knowing this, compare the seller's asking price to the prices being asked by reputable commercial seedsmen who will guarantee good quality seed with an extremely high live seed content per pound. Most farmers plant seed from their own harvested crops or which they buy from other farmers in the area. You would be unwise to do this if the seed sample test indicates you would be planting weeds, extraneous crops, and a whole drill box full of troubles.

Watch out for "Seedy" Operators

There are all sorts of bunco artists going around posing as legitimate seedsmen. Each year these gyppers seek out new and inexperienced farmers like you. They will tell you about a "great buy" their company was able to make of a highly desirable variety of seed. Or they may say they are offering select farmers a chance to plant "foundation seed" of a new variety not yet in the hands of other seed companies. Don't jump at a seedsman's offer of a "steal" until you have checked him out thoroughly. And don't just rely on your neighbors to spot these "seedy" characters; they may be duped right along with you.

The USDA and the various state departments of agriculture should be able to verify the legitimacy of a seed company representative. Be careful or that "steal" may turn out to be *your money!*

Another thing to be concerned about is the possibility of seed-borne disease organisms being present in the seed you buy. Some of these diseases will not only spoil this year's crops but they may infest the soil and cause damage for years to come. By purchasing certified seed you will insure that your seed is disease-free.

Write to the major seed companies requesting information about the type of planting you intend to do and about the varieties of seed that grow best and get the highest yields in your area. It is quite likely that they will furnish this information gratis and send out a company representative to give you on-the-spot consultation. Remember, you are a potential customer for years to come.

The seedsman will tell you, as the SCS man did earlier, that you should initiate a good crop rotation program if you want to get the highest seed-grain yields from your cropping. How many years in succession you may expect to plant a field with the same seed grain depends on the quality of your soil and the quality of seed you wish to harvest. The SCS and your seedsman will help you determine this.

Cereal grains like wheat, oats, barley, and rice will require the use of a seed drill for planting. The drill will give you a uniform planting at the proper depth. Drills should be maintained in good condition and be cleaned with disinfectant after each use.

Selecting the Tools of Your Trade

Recently I was watching the evening news on television and saw a filmed report on the spring floods that are causing devastation in the southeastern part of the country. The film showed long lines of farm machines crawling slowly along the highways—miles and miles of harvesters—heading for high ground. To the average nonfarming viewer, it must have looked ludicrous to see so many trying desperately to save their farm machines instead of concentrating their efforts on saving their homes and household effects. But any experienced farmer would have sympathized. He would know that farm machinery and the costs of owning and operating it are often the highest investments a man must make in order to become a successful farmer. It's been said that farming requires a greater investment in machines per man than any other industry. *The machinery on the farm you buy may cost as high as 40 percent of your total investment.* It is of critical importance, then, that you make every effort to keep your machinery costs within workable limits.

Often, the new farmer will have no real idea of what farm machinery he needs or what his tool and machinery costs should be. He is shown a couple of sheds and garages full of equipment that the seller has accumulated and he naturally assumes that these are all necessary items. That may not be true. Research from the Georgia Experiment Station has shown that machinery costs are higher than necessary on many Georgia farms. New farmers are especially prone to becoming overstocked with machinery that will not pay its way. There is little reason to doubt that this is the case on farms all over the country.

You, most certainly, will not want to spend more on farm machinery than you could possibly expect to pay for out-of-farm earnings. Nor would you want to pay for a whole lot of equipment that may not be useful or necessary to the success of your intended farm operation. Obviously, this is another one of those times for careful planning.

One of the first things you should do is to get your county farm advisor to help you prepare a list of the tools and machinery you will need to get started. This will be your basic inventory list. Just in case he is related to John Deere, Henry Ford, or the people who own Massey-Ferguson, you might be wise to have some other experts check the list over and make suggestions. Check your list with the state cooperative extension service and with other farmers who have successful operations of the type you intend to manage. These experts will be able to help you formulate a basic machinery and tool inventory which will be listed in the order of your most urgent priorities. In other words, *buy the tools you need most first, then have a long-range plan of acquisition for the other labor-saving and income-increasing equipment that you eventually hope to buy.* This second part of your list should be purchased on a pay-as-you-go basis. Try to overcome the urge to splurge and buy everything at once. Or you may find out to your chagrin that your new binder has put you in a bind!

Now that you have a reasonable shopping list of the equipment you are going to need, you can go ahead and make up a similar list of the machinery and tools that the present owner wants to include in the sale. Immediately eliminate expensive equipment that is not on your shopping list. Obviously, the equipment being offered for sale will vary from farm to farm according to the type of operation, the size of the farm, the layout of the fields, the financial resources, and the abilities of the individual operator.

Find out the age and condition of each piece of equipment. As you look at different farms, you will find that some of the tools and equipment have been in use for as much as fifteen to twenty years. Make certain this equipment is not obsolete. Sometimes you will find relatively new equipment that is obsolete. For example, if a farmer buys a new cotton harvester and later decides to convert his croplands to corn, the cotton machine becomes obsolete because it is no longer useful and necessary in his farming operation.

Try to decide how well the machinery on the farm matches your shopping list and how well it fits into your long-range farming plans. If you have intentions of planting orchards where vegetables were grown before or if you have plans to change the layout of the fields in order to be able to use large, more efficient, labor-saving equipment, it may be that you will not want very much of the gear on the seller's list.

Sometimes the seller is not getting out of farming. It may be that he is trading up to a larger farm in another location. This happened recently to the Wilbur Hageman family who sold their farm outside of Chicago and purchased a farm in western Minnesota. Will had no reason to sell most of his tools and equipment because he needed them on his new farm. All the heavy machinery was moved to Minnesota by truck and rail. If you are going to buy such a tool-less farm, you will have to buy almost all your farm equipment elsewhere. This does not necessarily mean that you will have to purchase all new equipment.

If your funds are limited, you may want to investigate farm auctions. It often happens that when a farm is sold, it is bought by a neighboring farmer who wants to increase the size of his holdings. These buyers have no need for the tools and farm machinery of the seller. You can check out these farm sales for great buys in used tools and farm machinery on your shopping list. Just be careful not to let the thrill of auction bidding cloud your basic common sense to the point that you buy something that's not on your list or pay more for things than they are really worth.

Another potential way to cut your farming costs is to look into the possibilities of *custom hiring*. Many times neighboring farmers will hire out to plow, disk, spread lime, cultivate, or harvest your crop. There are many advantages to this system if you are going to farm on a small scale. The Georgia Cooperative Extension bulletin, "Farm Machinery Cost Analysis," lists the advantages and disadvantages of custom hiring as follows:

Advantages

1. All fixed or ownership costs would be eliminated.
2. Capital that would be invested in machinery could be used for other enterprises.
3. You benefit from newer, more specialized machinery.
4. Annual machinery costs can be varied to suit the weather.

Disadvantages

1. The custom service may not be available when the job is ready.
2. The quality of custom work may be low.
3. Labor released by hiring custom service may not be put to other productive uses.
4. Diseases and noxious weeds may be spread from one farm to another by equipment.

Other alternatives include leasing farm equipment, renting it from a neighbor who is not using it, or joining a cooperative and sharing the costs of ownership with other farmers.

A thorough understanding of your equipment needs and the costs you are likely to incur will help you make the right decisions in selecting the machinery for your farm. Write to your state extension service for information on cost analysis. Or write to the Director, Cooperative Extension Service, University of Georgia, College of Agriculture, Athens 30601, and ask for Bulletin #637. This excellent and informative guide lists the following tips for cutting your farm machinery costs:

1. Figure the cost of owning and operating different sizes of equipment.
2. Select the proper size equipment for your farm needs.
3. Compare the cost of owning a machine and custom hiring.

4. Figure the economics of a trade before you make it.
5. Shop around for the best buys.
6. Extend useful life of equipment through proper operation and preventive maintenance.
7. Keep machinery properly adjusted.
8. Have repair work done in off-seasons.
9. Reduce your investment in machinery by joint ownership of some machines.
10. Keep complete records on the cost of operating the equipment for future reference in making machinery decisions.

Also, when it comes to farm equipment, forget the old adage, "Neither a borrower nor a lender be." Instead, be a good neighbor.

Section Five
Getting Your Farm

Ilene's Grandpa Richmond used to say, "There aren't many problems on this farm a little money would make any worse." Grandpa sure had something there! One of the greatest assets you can bring to your new farm enterprise is money. You probably didn't have to look at too many farms to realize you will need large amounts of money to pay for the land, the farm buildings, the livestock, seed, fertilizer, machinery, the operating capital, and your country home. Although you may be willing, able—even *eager*—to take on the responsibilities of owning and managing a farm, your ability to get credit and financing will be one of the toughest parts of making your dream of a life in the country come true.

Finding a Friendly Lender

As soon as you pinpoint the county or country area where you want to locate your farm and start to survey potential farm buys, you should also begin to investigate the banks and other potential sources for obtaining credit. Even if you are unlike the vast majority of new farmers and have enough cash to buy your farm outright, you will still need credit for financing other farm resource purchases, improvements, and other farm operations.

Too many folks just don't realize that they should shop around for credit just as they would shop around for a farm or a new car. You should keep in mind that money is not much good to the people who have it unless they can put it to some advantageous use, just as your ability to build a successful farm operation is not much good unless you can put it to advantageous use. The lender, or lending institution, needs people to invest their money with, just as much as you need their money to purchase the useable resources or commodities that will insure your farm success. Lending institutions are looking for reliable people who will be able to give them a good return on the money they invest. This return comes to them in the form of interest and charges you pay on top of the principal when you repay the loan. You, of course, are looking for a lender who will have the utmost confidence in your abilities to repay his loan out of your projected farm income. You will naturally try to find a lender or lending institution who will not be trying to squeeze you for the highest interest, or the greatest amount of profit, he can get. In times like these, when money is in short supply, interest rates will be high. In times when money is readily available, interest rates will be more reasonable. You would be wise not to accept the going rate of interest until you have shopped around to make sure you are obtaining the best financing for the lowest loan charges and the least amount of interest you can find.

Before you consider the various sources for farm and rural credit, it will be good to have a basic understanding of the different methods of financing and types of loans these sources use.

If you intend to buy your farm, you will be seeking a "mortgage loan." The word "mortgage" is the term accepted by custom in this country for *a money encumbrance secured by real property*. The term mortgage is still used interchangeably in those states which use *trust deeds* for such encumbrances.

In the parlance of bankers and other lenders, there is a "primary" and a "secondary" mortgage market. Primary financing refers to the first or "prime" loan on a property, such as a first mortgage or first trust deed. Secondary financing refers to refinancing an existing mortgage loan or obtaining a second mortgage or trust deed. Primary financing sources are those lenders who supply funds directly to borrowers. They are lenders who take on the risk that goes with long-term financing and who usually hold the mortgage until the debt is paid. Secondary financing sources are lenders who sell or make loans on existing mortgages.

The most common type of loan you can obtain is called a conventional loan, which is one that is not insured by any government agency. Therefore, with conventional loans, you take on all the responsibilities of repayment and the lender assumes all the risk. The more money you put down, the less risk for the lender, and the less he is likely to charge for interest.

Your county agent can tell you which institutions in the area are most commonly used by farm buyers and established farmers to obtain conventional, insured, or guaranteed loans. These usually are: individual lenders, commercial banks, insurance companies, savings and loan associations, the federal land banks and their associations (FLBA), the farmers home administration (FHA), federal intermediate credit banks (FICB) and production credit associations (PCA).

The most common loans made to beginning farmers, a people new to the country, are for real estate, machinery, and livestock. Other operating expenses, like feed, fertilizer, farm labor, and gasoline, are also included in this category. These typical loans are called production loans. Other loan categories are: consumer, convenience, and emergency loans. If you plan your long-range credit needs carefully, you will see that obtaining initial money for production purposes will give you income and therefore will be the easiest type of loan to repay. Also, if you try to limit the majority of your loan requests to purposes which will produce income, you should soon be able to build up your capital reserves to the point where you may not have to borrow heavily for consumer, convenience, or emergency needs.

In the past, farmers were often reluctant to use much credit in their farm operations. Many farms were run on a pay-as-you-go basis. This resulted in the farm family fluctuating from "high-on-the-hog" living conditions to periods where everything seemed "lower than a sow's belly."

Today it is considered just as proper for you to use your credit to improve your farm operation as it is for industrial giants like General Motors to borrow money to improve their plants and keep them operating more efficiently. Too many beginning farmers don't have a plan, so they don't have confidence in their ability to use credit wisely. Sometimes this lack of confidence and fear of using credit will inhibit the success of their farm enterprise.

I'm not recommending that you allow your family farm business to drift heavily into debt because of deficit spending, but I am saying, don't be timid. Plan your credit needs carefully for the long haul and proceed with confidence.

Loan Sources

Borrowing Can Be "Relative"

Don't overlook your relatives or other individuals with capital when you are seeking the sources to lend you the money to help you get started in farming. Most of the research I have seen indicates that loans from individuals are the primary source of capital for families starting out in farming or country living. Who more than your relatives and friends knows how reliable you are? Where else can such individuals get a return on their investment capital and with it the satisfaction of helping someone succeed? You may find that such individuals will be the easiest to convince that you can make it as a farmer or country businessman. However, to avoid quarrels and family feuds which may crop up later, have a lawyer draw up all the necessary and proper contracts, spelling out the terms and conditions of your loan agreement. You should do this even if the money comes from doting parents. Each party should know, exactly, what he got and what he's got coming. In the long run everyone will appreciate having it down in black and white.

Sometimes family loans, or loans you get from individuals, are open-ended. That is, they don't have the term for repaying the loan spelled out. While this might seem like a good deal at first, I believe that it is better to have a specified date saying when the loan must be repaid. This will force you to set about meeting your financial obligation in a responsible and systematic fashion.

Sometimes relatives or friends who lend money to you to buy

a farm or purchase seed for a crop ask for repayment in cash. Other times they prefer a share of the crop or a piece of the farm. If that is the agreement, have your lawyer make certain you are identified in the contract as the sole farm manager and operator. Otherwise, you may jeopardize your tax status as a self-employed farmer. Such share-of-ownership relationships can prove to be problems, even when they are only in existence during the term of the loan. Make sure you would keep the farm if you got laid up by an accident and make sure the ownership of the farm reverts to you alone when the loan has been repaid.

Make the Local Banker Your Buddy

One nice thing about moving to the country is that people always seem to have more time for each other. Back in town, you or he may not have had the time to notice, but . . . bankers are people! I'd like to suggest that you take note of this fact and do something with it. Stop in at the local bank, in the area where you might buy a farm, and introduce yourself. Tell the banker, or loan officer of the bank, about your plans to settle on a farm near his community. Tell him about the farms you have been looking at and ask if the bank has financed any of them in the past. "One test of a farm's potential value is whether a bank is anxious to make a loan on its purchase or improvement." *

Tell the banker about your long-range farming plans and what you think your credit needs will be over the next several years. If you show him you like to plan ahead and are willing to listen to his comments and advice, he will quickly realize that you can become a good customer of his for many years to come.

It's important that you realize that your potential patronage in the town nearest the farm where you locate is a valuable asset. Any business, including a bank, should be willing to give friendly personal service. You will not want to do business with a bank run by people who are cold and unfeeling.

---

* Edward A. Yeary, "So You Want a California Farm," Univ. of California Cir. #556.

As you go shopping for credit, you should have your money requirements all worked out. You won't look like a very responsible farmer-businessman if you borrow money for real estate, then come back at a later date when you discover you need to make improvements on the farm home, then come back a month or two later when you decide you need a loan for seed and fertilizer, etc. If you have a plan which outlines what you think you are going to need, how you are going to use the money, and how you project you will be able to repay it, the loan officer or banker will be able to examine your qualifications, weigh the risks, and make a speedy decision. In most cases, that decision will be favorable.

While on a national scale commercial banks are not the most important holders of farm mortgages, you may find certain communities where the local bank is the *only* easily available mortgage lender. If this is the case, see if you can obtain mortgage money for a reasonable amount of interest payments. Right now, interest rates vary between 9 and 11.5 percent, except in states where there is a legal maximum limit. Banks will often charge at the high end of the going interest rates unless you show you are willing to share their risk with a hefty down payment. They may also have the strictest eligibility requirements and the shortest-term loans, sometimes no longer than ten years.

Commercial banks usually want a down payment of 20 to 50 percent on farm mortgages, depending on how you measure up to their other eligibility criteria. In a recent study of bank loans to beginning farmers in Alabama (Alabama FB-400), the common repayment schedule was: nine to ten years for real estate loans, two to three years for machinery and livestock loans. If you make a good impression on your banker, he may go out of his way to arrange a longer-term loan through an insurance company or some government agency.

Make a list of your business and personal assets. These are: the down payment or collateral you have to put up as security on the loan, your good financial reputation, your previous experience in farming or a farm-related business, your willingness to accept technical assistance planning from a state or federal agency, your formal education, military service record, and your projected ability to repay the loan.

Since you may not have a great deal of capital to use as down payment and collateral, you should play up your other, stronger areas of eligibility. Don't discount the importance of your formal education or your military service record as eligibility criteria. Statistics have shown that people who complete twelve or more years of schooling will earn somewhere around half a million dollars during their working lifetime. And people who have served their country honorably and responsibly are very likely to conduct themselves in the same way when it comes to handling their financial obligations.

If you are a veteran, the bank may be able to arrange a mortgage or loan insured by the FHA or VA under the provisions of the G.I. Bill. One disadvantage to these loans is that the G.I. Bill only insures loans up to the amount of $12,000, which is not enough mortgage money for most farms. Also, you may have difficulty finding a lending institution willing to make G.I. loans because of the low interest.

If you are a little light on previous farming experience, you might want to show the banker that you are going to follow a detailed farming program you have planned for your farm with the aid of the state cooperative extension service, the SCS, or some other agency which offers technical assistance, like the Farmers Home Administration. Another asset which may help you obtain a mortgage loan is a credit life insurance policy which guarantees repayment in full in case of your death.

Because commercial bank mortgages are usually short-term, high-interest loans made for relatively small amounts, you might be wise to ask your banker to help you apply with some of the other lending institutions in the area which offer lower interest payments or long-term financing.

Loans from Insurance Companies

Many life insurance companies invest a large proportion of their premium capital into farm mortgages. These companies are a good source to apply to for a farm mortgage loan. The terms will probably be more lenient and the interest charges more reasonable than your banker can give you. Usually, the loans are made for twenty to twenty-five years. Interest will be on the low end of the going rate and you may be able to get them even lower if you will take out a credit life insurance policy with the company. Often, farmers are allowed to make prepayments without penalty and to skip payments when droughts or emergencies limit farm income.

Mortgages from Savings and Loan Associations

Savings and loan associations and the mutual savings banks, which are their counterparts in the northeastern part of the country, are not usually involved in the farm mortgage business. However, if you buy your farm as a second home or a vacation home while continuing to maintain a city residence, you may want to contact your savings and loan association to see if mortgage money is available. They usually make loans for longer terms at the going interest rate.

Federal Land Bank Loans

There are twelve federal land banks which have been set up under the supervision of the Farm Credit Administration (FCA) of the USDA. The land banks make long-term loans of five to forty years. Those who may be eligible for the loans include farmers, growers, ranchers, certain corporations engaged in agricultural production, rural residents, and selected farm-related businesses. You may apply for a loan at the local Federal Land Bank Association (FLBA) serving the area in which the farm property you want to buy is located. The loan, if it is approved, will be secured by a first lien on real estate held by the local association. There are more than five hundred of these local federal land bank associations in the territories served by the twelve federal land banks.

According to FCA Circular #35, "Federal Land Banks & How They Operate," long-term loans may be obtained for any agricultural purpose. These include: purchase of farms, farmland, equipment, and livestock; refinancing existing mortgages and paying other debts; constructing and repairing buildings; financing other farm, farm home, or family needs. Owners of farm-related businesses may borrow to purchase or refinance necessary sites, capital structures, equipment, and for initial working capital. Loans for rural residents may be obtained for building, buying, remodeling, improving, refinancing, or repairing a rural home.

When you apply for your land bank loan, you will be required to give information regarding your financial status, the purposes for which you want the loan (what the money will be used for), and other facts pertinent to the loan. The FLBA will then send an appraiser out to look at the property which is being offered as security while your credit worthiness is being checked out.

Down payment requirements for land bank loans usually range somewhere between 20 to 50 percent of the appraised value of the property. Within limits, repayment plans can be geared to your individual preference or business setup. Payments may be monthly, quarterly, semiannual, or annual. As with other bank loans, the installments include accrued interest on the unpaid balance and sufficient principal payments to retire the loan in the number of years of the term for which it is made. A recent plan, introduced by the FLBAs, allows for you to defer the principal payments for as long as the first five years.* This deferment plan allows you to invest your income capital toward expanding your farm and putting it on a firmer commercial foundation without having to pay on the real estate principal.

If you are approved to borrow through the local FLBA, you will be required to purchase stock in the association in proportion to the amount of your loan. Stock dividends which accrue may be used to build your capital reserves or toward the early retirement

* "Financing Michigan Farms—Now and in 1980," J. R. Brake, Mich. Agr. Exp. Stn. and Coop. Extn. Svc. Resources Report #46, 1966.

of your loan indebtedness. As a stockholder, you will have one vote in electing the board of directors of your association. It is the job of the board to see that all applicants are treated fairly and all loans are impartially and properly serviced.

FEDERAL LAND BANKS AND TERRITORY SERVED BY EACH *

The Federal Land Bank of
Springfield
P.O. Box 141
Springfield, Massachusetts
01101
Maine, New Hampshire, Vermont, Massachusetts, Rhode Island, Connecticut, New York, and New Jersey

The Federal Land Bank of
Baltimore
P.O. Box 1555
Baltimore, Maryland 21203
Pennsylvania, Delaware, Maryland, Virginia, West Virginia, District of Columbia, and Puerto Rico

The Federal Land Bank of
Columbia
P.O. Box 1499
Columbia, South Carolina
29202
North Carolina, South Carolina, Georgia, and Florida

The Federal Land Bank of
St. Paul
375 Jackson Street
St. Paul, Minnesota 55101
Michigan, Wisconsin, Minnesota, and North Dakota

The Federal Land Bank of
Omaha
P.O. Box 1242
Omaha, Nebraska 68101
Iowa, Nebraska, South Dakota, and Wyoming

The Federal Land Bank of
Wichita
151 North Main
Wichita, Kansas 67202
Oklahoma, Kansas, Colorado, and New Mexico

The Federal Land Bank of
Houston
P.O. Box 2649
Houston, Texas 77001
Texas

* Courtesy of FCA. For more information and loan application forms, contact your local FLBA, or The Farm Credit Administration, 490 L'enfant Plaza, S.W., Washington, D.C. 20578.

The Federal Land Bank of
Louisville
P.O. Box 239
Louisville, Kentucky 40201
Ohio, Indiana, Kentucky, and
Tennessee

The Federal Land Bank of
Berkeley
P.O. Box 525
Berkeley, California 94701
California, Nevada, Utah,
Arizona, and Hawaii

The Federal Land Bank of
New Orleans
P.O. Box 50590
New Orleans, Louisiana 70150
Alabama, Mississippi, and
Louisiana

The Federal Land Bank of
Spokane
W. 705 First Avenue
Spokane, Washington 99204
Washington, Oregon,
Montana, Idaho, and Alaska

The Federal Land Bank of
St. Louis
Main P.O. Box 491
St. Louis, Missouri 63166
Illinois, Missouri, and Arkansas

FHA Farm Loans

If you are finding it hard to obtain credit from the sources I have mentioned so far, your banker may suggest placing an application with the Farmers Home Administration of the USDA, which was created to channel credit to farmers, rural residents, and communities of less than ten thousand inhabitants. In cases where a borrower is having difficulty finding a lending institution which will furnish credit at a reasonable rate of interest or in the amounts needed, the FHA will help eligible applicants obtain a guaranteed or insured loan.

Guaranteed loans are made and serviced by a private lender, but the FHA guarantees that it will limit any loss to the lender up to a specified percentage of the amount loaned. Insured loans are made and serviced by the FHA. Interest rates for these two loan categories are either agreed upon between the lender and borrower or they are determined by the current cost of federal borrowing. Exceptions to these practices are made when the rates are established by federal statute.

You can apply for FHA-insured loans at whichever of the forty-two state offices is closest to the locality where you intend to locate your farm, rural home, or farm-related business. Guaranteed loans may be applied for through your bank or an FHA-approved lending institution.

The three types of FHA loans most likely to interest you will be: farm ownership loans, farm operating loans, and farm emergency loans.

According to USDA-FHA Program Aid #62, FHA farm ownership loans may be obtained for the following purposes:

1. To purchase a farm
2. To enlarge a farm and make it more economically efficient
3. To build, enlarge, or remodel a farm home
4. To build, enlarge, remodel, or repair farm service buildings
5. To drill wells, improve water supplies
6. To refinance debts
7. To develop and improve farmland
8. To clear and level land
9. To establish and improve farm forests
10. To provide drainage systems
11. To carry out basic land treatment practices such as liming and make other improvements
12. To construct or provide facilities to produce fish under controlled conditions
13. To finance nonfarm enterprises to help farmers supplement their farm incomes—buy and develop land, construct buildings and other facilities, purchase equipment, and make other real estate improvements
14. To finance nonfarm enterprises on family farms like camping and swimming facilities, tennis courts, riding

stables, vacation rental cottages, lakes and ponds for boating and fishing, docks, nature trails, picnic grounds, repair shops, roadside markets, souvenir shops, craft and wood- or metal-working facilities, and small grocery stores or service station facilities

The terms and interest rates are as follows:

1. Loans may be granted for a maximum of forty years.
2. The interest rate is 5 percent per year on the unpaid principal.
3. The borrower may make large payments in years of high income with no penalty. This will build up a reserve that may keep the loan in good standing during years of low income.
4. The borrower is expected to refinance the unpaid balance of the loan when it is financially feasible for him to rely solely on commercial credit sources.
5. Farm ownership borrowers are required to maintain their property and pay taxes and property insurance premiums when due.
6. The loan may not exceed the market value of the farm minus any liens against the property that will remain outstanding, and may not exceed the amount certified by the country FHA eligibility committee. An appraisal will be made of the security property to determine its value.
7. The maximum farm ownership loan principal may not exceed $100,000 and the combination of all debts against the security property may not exceed $225,000.

According to USDA-FHA Program Aid #1002, FHA farm operating loans may be obtained for the following purposes:

1. To buy cattle, hogs, other livestock
2. To buy poultry
3. To buy tractors, plows, sprayers, other farm equipment
4. To buy freezers, other home equipment

5. To buy fencing
6. To make minor improvements to buildings and land
7. To develop water supply systems for home use, livestock, and irrigation
8. To buy feed, seed, tractor fuel, lime, fertilizer, chemical sprays, hail and crop insurance, and hiring labor
9. To purchase and pay for food, clothing, medical care, and personal insurance
10. To buy equipment and to pay operating expenses for producing and harvesting trees and other forestry products
11. To finance the production of fish under controlled conditions in ponds, streams, and lakes
12. To finance nonfarm and recreational enterprises
13. To purchase equipment, animals, and game birds
14. To finance minor real estate improvements
15. To pay expenses of operating fishing, horseback riding, camping, hunting, machine shop, and other enterprises
16. To refinance debts other than real estate debts
17. To purchase membership and stock in farm purchasing and marketing and service-type cooperatives and certain types of recreational cooperatives.

The terms and interest rates are as follows:

1. Loan funds may only be used to pay for those items essential to the success of the proposed operations.
2. Interest rate is set each July 1. Repayment is scheduled according to the borrower's ability to repay. Funds advanced for operating expenses are repaid when the crops, livestock, or other produce are sold.
3. Funds advanced for other purposes may be repaid from one to seven years; in some cases, loans may be renewed for up to five years.
4. A borrower may make large payments in years of high in-

come to build a reserve that keeps the loan in good standing during years of low income.

5. Each borrower must refinance the unpaid balance of the loan when able to obtain such refinancing at reasonable rates and terms from other lenders.

According to USDA-FHA Program Aid #973, FHA farm emergency loans may be obtained for the following purposes:

1. For operating or living expenses
2. For home repairs
3. To replace livestock, essential farm buildings, and equipment needed to restore normal operations after severe loss from natural disaster in designated areas
4. Some debts may be refinanced

To be eligible for any of the three types of FHA loans, you must:

1. Have recent farm experience or training needed to succeed in the farming operation
2. Possess the character, industry, and ability to carry out the farm operation or the nonfarm enterprise
3. Manage or operate the farm or nonfarm enterprise
4. Be unable to obtain sufficient credit elsewhere at reasonable rates and terms to finance your actual needs
5. Be a citizen of the United States and of legal age
6. After the loan is made, you must be the owner-operator of a family farm that will produce a substantial portion of your total income
7. Be able to obtain operating capital, including livestock and equipment
8. Be an individual who must rely on farm income to have a reasonable standard of living

Veterans of military service are automatically given preference. All applications are considered regardless of race, color, creed, or national origin of the applicant. If you feel you may qualify, you should apply at a commercial bank or savings and loan or at the local FHA office serving the area where the property is located. All rural counties are served from over 1,700 offices, usually located in the county seat. If you can't find the FHA listed in your telephone directory, write to Farmers Home Administration, U.S. Department of Agriculture, Washington, D.C. 20250, for the address of the FHA office in your locality.

With all FHA loans, a committee, composed of three persons who know the farming conditions and the credit conditions of the county, or area, screens the applications. Often, the committee asks the farmer and his wife to meet with them, so be prepared for that eventuality.

If you are eligible, the FHA will supply an expert to help you work out a detailed plan for getting the most productive use out of your land, business, labor, livestock, capital, and equipment. The borrower is expected to keep accurate records showing he is following the technical assistance program.

Federally Funded Credit Sources for Shorter-term Loans

It may be that you will not need long-term financing in order to purchase or operate your farm. In that case, you may want to apply for a short-term, or intermediate-term, loan at the Production Credit Association (PCA) office in your local area.

There are more than four hundred production credit associations across the country which make loans to farmers, growers, ranchers, producers, and harvesters of aquatic products, rural homeowners, and farm-related businesses. These loans may be made to you with funds provided the PCA by the Federal Intermediate Credit Bank (FICB) serving the territory where the farm, rural home, or farm-related business is located.

In some cases, the FICBs may participate with PCAs in making loans. Or the FICB may discount the notes of eligible borrowers given to other commercial banks or lending institutions.

According to FCA Circular #36, "The Cooperative Farm Credit System," loans may be obtained for purposes related to: the purchase, repair, or maintenance of rural homes; the production of agricultural products; the production and harvesting of aquatic products; and other requirements of the borrower.

The terms of PCA loans are for periods of up to seven years. And the amount of money you may borrow cannot be more than 85 percent of the value of the assets devoted to your agricultural or aquatic enterprise.

To apply for a loan, contact the PCA serving your county or rural area and outline your reasons for borrowing and plans for using the money. If your plan sounds worthy, the PCA official will ask you to fill out a formal application and furnish the PCA with a financial statement. Together, you and the PCA manager or field representative will go over your operation, determine how much money you need, and work out a repayment plan. A PCA representative will then look at your farm operation to help you decide the best course of action. He will help you develop a long-range credit plan so you will be able to draw money from the PCA as you need it in your operation.

When your loan goes through, you will be required to purchase stock in the PCA in proportion to the amount of the loan. Each successive time you borrow, you will have to buy more stock. Stock dividends and patronage refunds will be paid to you at given intervals. As one of the owners of your PCA, you will have some control and voice in the processing and servicing of credit dealings. As with all loans supervised by the FCA or federal government, there is no penalty for prepayment.

For more information on short-term and intermediate-term loans, contact your local PCA or The Farm Credit Administration, 490 L'enfant Plaza, S.W., Washington, D.C. 20578, or, the federal farm credit bank serving your territory.

FARM CREDIT BANKS AND TERRITORY SERVED BY EACH *

The Farm Credit Banks of Springfield
Springfield, Massachusetts 01101
Maine, New Hampshire, Vermont, Massachusetts, Rhode Island, Connecticut, New York, and New Jersey

The Farm Credit Banks of Baltimore
Baltimore, Maryland 21203
Pennsylvania, Delaware, Maryland, Virginia, West Virginia, District of Columbia, and Puerto Rico

The Farm Credit Banks of Columbia
Columbia, South Carolina 29202
North Carolina, South Carolina, Georgia, and Florida

The Farm Credit Banks of Louisville
Louisville, Kentucky 40201
Ohio, Indiana, Kentucky, and Tennessee

The Farm Credit Banks of St. Paul
St. Paul, Minnesota 55101
Michigan, Wisconsin, Minnesota, and North Dakota

The Farm Credit Banks of Omaha
Omaha, Nebraska 68101
Iowa, Nebraska, South Dakota, and Wyoming

The Farm Credit Banks of Wichita
Wichita, Kansas 67202
Oklahoma, Kansas, Colorado, and New Mexico

The Farm Credit Banks of Houston
Houston, Texas 77002
Texas

The Farm Credit Banks of Berkeley
Berkeley, California 94701
California, Nevada, Utah, Arizona, and Hawaii

* Courtesy the FCA

The Farm Credit Banks of New Orleans
New Orleans, Louisiana 70130
Alabama, Mississippi, and Louisiana

The Farm Credit Banks of St. Louis
St. Louis, Missouri 63166
Illinois, Missouri, and Arkansas

The Farm Credit Banks of Spokane
Spokane, Washington 99204
Washington, Oregon, Montana, Idaho, and Alaska

Central Bank for Cooperatives
Denver, Colorado 80202
Serves District Banks for Cooperatives

Buying Your Farm

If you have found a way to obtain credit and made a careful study and assessment of all the features of the farms you have been inspecting, both good and bad, you are now in an excellent position to pick out the farm that scored the highest in your estimation and begin negotiating for its purchase.

Before You Begin Negotiations . . .

There are a few more things you should know about the property in question before you make an offer to buy it. You can find out these things by going to the office of the county recorder and checking for encumbrances, taxes, and assessments which may affect your decision to purchase the property. An encumbrance can be defined as anything which affects the title to or limits the use of real property. They are generally classed as *money encumbrances* or *nonmoney encumbrances.*

Money encumbrances are called *liens.* Liens are charges against the property owner, making the property security. Some liens are *voluntary,* where the owner has willingly used the property as security to borrow money. Others are *involuntary,* liens placed on the property against the expressed will or consent of the owner. For example, when the owner fails to pay his property taxes and the state puts an involuntary lien on the property. Other examples of involuntary liens are charges such as special assessments, personal property taxes, mechanic's liens, judgments, and attachments.

Nonmoney encumbrances are often referred to as *claims*. They are the covenants, conditions, restrictions, reservations, rights, rights-of-way, and easements (C,C & Rs) I mentioned earlier in the book. An example of a convenant might be when property owners get together and agree to use or not use their land in some way—perhaps limiting an area of country property to camping sites. Obviously, you would not want to buy such property and find you were legally restricted from building a farm or country home. There are an almost unlimited number of possible C,C & Rs, so it will be well worth your while to check the deed for them.

Zoning is an example of public control or restrictions which may be placed on real property. You should check the zoning controls to see how they affect *land use, limits on structures* in height, bulk, size, and the number of families that may reside in them, and *lot areas*. Obviously, you cannot farm where the zoning forbids land use for agricultural purposes. Other examples of public control and restrictions are health, safety, and building codes.

Taxes

Property taxes will not be recorded with the deed, but they are liens on the property just the same. The tax rate is based on a certain number of dollars per assessed value of the property. The assessment is usually just a small percentage (depends on what you call small) of the market value of the property. Often, farms and country homes will not have been assessed for years. Because of this, the tax rate may be very low and you may be getting a farm in what might be called a "pocket tax paradise." Unfortunately, after you buy, you may quickly discover that your tour in "heaven" is going to be short-lived. Most country tax assessors have people, known as observers, whose job it is to regularly check the deeds in the county recorder's office. When one of these observers finds that a piece of property has been recently sold, he checks the number of tax stamps on the deed. These stamps will indicate to him if the property was sold for a greatly increased price. If it was, he will notify the assessor, and you can be sure that your next tax bill will be increased as the result of a "reassessment."

Taxes on land are usually figured on its "highest and best use." If the property you are looking at is located near a large city or a growing community, this "best use" will be determined by what a developer would do with it. In that case, the land may be taxed at too high a rate for you to farm it economically.

If you want to check further about the taxes and tax trends on the property, you can go to the county assessor's office and ask for the tax records of the land that include the property you are interested in. The clerks there will be able to interpret the information and tell you about any special assessments that have, or are, going to be made. Or about any reassessments being planned for the whole area which might result in a raising of the taxes. If you find that betterments and special assessments have been made recently and the present owner has not yet paid his portion, you will want to consult your realtor and attorney about ways to get him to pay the new taxes, or, at least, share the payments with you on a prorated basis for the year.

Real estate taxes have always been a burden to farmers, but in recent years, the tax weight has shown signs of becoming unbearable. At the turn of the century, the national average farm property tax was 13¢ an acre and totaled $105.6 million. Over the years, farm taxes have risen steadily. But since 1942, when they were 38¢ an acre and $399.5 million, they have gone up every year at an accelerated rate of increase. Nineteen hundred and seventy-three was the most recent year surveyed by the Economic Research Service of the USDA. The ERS says that in that year farm taxes reached a record $2.45 billion! The new national average tax was $2.56 per acre, an increase of 2.5 percent over the previous year. Since the property tax is the main source of revenue for most state and local governments, the ERS concluded that there is little hope, at present, that farm taxes will go down or stabilize.

Unfortunately for you, as a beginning farmer, the prices of farm products and commodities have not risen in proportion to taxes or inflation. A continuing of this trend of accelerated increases in farm property taxes is bound to be an important factor to the success and survival of your farm. Later on you may wish to join the UFO, NFO, Grange, or Farm Bureau and become an active advocate of tax reform measures for agricultural land.

Selling Development Rights

In a *Time* magazine article entitled "Saving the Farms" (April 21, 1975), there is a report on a new twist which may help keep farmland taxes within bounds and help save rich agricultural land from being lost to developers. In Suffolk County, Long Island, New York, a scheme has been initiated which allows farmers to sell the development rights of their farmland to the county. In other words, the farmer gives up the right to sell his land to developers in return for several thousand dollars per acre. The county pays the farmer the difference between the high value the property would have to a developer and the low value it would have as farmland. As part of the program and in return for the money, the farmers must agree to having a covenant to their deed which insures that the land will remain as a farm forever.

According to the *Time* report, farmers benefit from the program in three ways: (1) they get a tax advantage because all taxes are determined on the land's assessment for agricultural use only, (2) they get large amounts of cash, which may be used for needed farm improvements or to buy costly new farm equipment, (3) they retain their pride in ownership and their stake in farming. Proponents of the program say that the general tax-paying public receives equal or greater benefits.

Although a program for selling development rights will only have an immediate bearing on your present farming plans and projected taxes, if you happen to be planning to buy a farm in Suffolk County, it is something you might want to know and think about for the future. If so, I suggest you contact Mr. John Klen, County Executive, Suffolk County, Long Island, New York.

Now, let's get back to the problem at hand. If you have found out all you can about the encumbrances, taxes, and assessments which relate to the property and have found no major obstacles, you are probably ready to make an offer to the seller.

Negotiating the Purchase

Unless you are buying a second home or country vacation spot, buying a farm will be one of the greatest capital investments of your lifetime. I strongly recommend that you hire an attorney to represent your interests in negotiating the purchase. He will be well worth the money he charges you, even if the seller offers to share an attorney with you and even though the bank will have a disinterested attorney who is supposed to look out for your interests as well as those of the bank and the seller. You and your attorney should take the time and the trouble to go over the purchase, step by step, to make sure you are fully protected, in writing.

The First Offer, Binder, and Deposit

Even though you already know what the seller is asking for the farm, you will want to contact the realtor and have him present your first offer with a reasonable deposit. If your attorney is someone who specializes in real estate transactions, he will be able to advise you on what a reasonable first offer and deposit might be. Since the realtor is really representing the seller, he is not necessarily the best source for advice on this amount. If the realtor says you must sign a written memorandum of the transaction, called a *binder*, don't do this without first going over its contents with your attorney.

Before you sign the binder, your attorney may suggest you have an experienced farm appraiser look at the property and that the property be surveyed. He may also insist that the seller furnish proof that the farmhouse has been inspected and found free of termite, insect, and fungi damage. If you agree to these suggestions, and you should, he will alter the binder to make it subject to your satisfaction with the results of an appraisal, survey, and termite inspection. If you are not absolutely sure you can obtain a mortgage at the percentage of interest you believe is fair, the binder should also have a contingency clause stating that your offer is contingent on obtaining a mortgage at the specified interest. Until you sign the binder, any offer you make may be withdrawn for any reason. Three days after you sign, you may withdraw the offer only for reasons stipulated in the contingency clauses you have added. If you withdraw from the purchase for any other reason, you may be obliged to forfeit the "earnest money" you have deposited.

Assuming that the farm appraiser turns in an itemized report that is in line with what you thought the land, farm home, farmstead features, livestock, seed, feed, fertilizer, machinery, and other items to be included in the sale are worth, and assuming that the appraisal is not too far below what the seller is asking for the property, you will want to continue the negotiations.

Assuming the surveyor's report shows the property to be substantially as the seller and deed describe it, assuming the termite inspector's report warrants that the house is free of infestations; also, assuming you can get the type of mortgage you specified in the binder, you will want to continue negotiating.

Bargaining

Having watched my Uncle Art, who owns Good Time Farms, buy and sell horses with professional horse traders, I have come to realize that bargaining is a business technique and an art. I'm afraid that most of us are poor horse traders. Some people have a tendency to drag their feet and be overcautious.

Often this causes them to lose the property they really want. Others, like me, are too abrupt, and want to get the thing over with. Usually we pay too high a price. If either of these two models fits you, it might be a sound idea to let your attorney do the bargaining for you in attempting to get the price of the property as low as you can. If you decide to do this, be sure to know in advance the parameters he is using to determine where he thinks the final price, acceptable to you and the seller, will end up.

If the seller turns down your first offer (as he probably will), he may at the same time offer a counterproposal. You and your lawyer will have to decide whether to accept this or continue bargaining.

If one of your offers is eventually marked "accepted" and signed by the seller and his wife and/or partners, you will have a legal document of sale. If he accepts your offer without signing, your lawyer will want to draw up a purchase contract or contract of sale. More than likely, this will be done with the combined efforts of both attorneys.

K.I.S.S. Your Purchase Contract!

My realtor friend, Col. Willy Williams, says that K.I.S.S. is an admonition that he and his colleagues apply to all contracts. K.I.S.S. means "Keep It Simple, Stupid!" Sometimes attorneys have a way of getting wordy and confusing what can and should be a simple transaction. The most important elements to be included are the legal names of the buyer and seller, the legal description of the property being sold, the method of payment, and the date of transfer. In addition to these key elements, contingency clauses may be added by either party to limit their obligations under special circumstances. The contingencies you might ask for as a buyer include those mentioned earlier in discussing the binder. In the case of farm property where livestock, crops, and machinery are also being transferred, a legal description of such property should also be included. Contingencies regarding the certification of dairy herds or other livestock or poultry might also be included. If your attorney has previously handled farm sale and purchase transactions, he will know how to draw up the purchase contract properly. Also, consult your county agent for advice on this aspect of the contract. The attorney for the lending institution will also take pains to see that the contract is correct.

As soon as you have signed the purchase agreement, you should call your insurance man and ask him to issue an insurance binder on the farm and all the other property you will receive in the sale. This will protect the interest in the property you have just acquired. The binder is a temporary type of policy. While it is not very specific, it will be adequate until after the closing.

The Right to Cancel

Federal law permits you to cancel and withdraw from the transaction even after you have signed the purchase contract, for any reason, if you do so within three business days. You must cancel by mail or telegram no later than midnight of the third business day. You may do this without incurring any penalties and will have your down payment returned. Also, the Truth in Lending Law allows you the right to cancel, at any time, even after the three-day period, if the seller or his agent has failed to disclose, or misrepresented, any negative information pertinent to the sale.

The Closing

The closing is usually held at the bank or lending institution holding the mortgage. The attorney for the bank will preside at this "settlement" of the loan. At the closing you will be responsible for certain charges including: drafting documents, notary fees, lender's title insurance policy, half the escrow fees, etc. Certain charges will be the responsibility of the seller. Certain items like taxes (paid or unpaid), interest, insurance, etc. will be divided between the buyer and seller and prorated.

When all the papers have been signed at the closing, your dream of owning a piece of country property will be realized! Or will it?

Land Contracts

If you are having trouble raising the down payment or meeting other mortgage-loan eligibility criteria, you may still be able to purchase the farm or country place you want through an agreement with the seller known as a land contract. These low-equity financing agreements are also known as conditional sales contracts, purchase contracts, and contracts for sale—depending on the common business usage in your locality.

With a land contract, you will gain possession of the property almost immediately, but the deed will remain with the seller until you have paid a specified amount to the seller in installments. Interest charges must be made on the unpaid balance just like on mortgage payments. With most installment-type land purchase contracts, the down payment required may be only about 10 to 20 percent of the total selling price instead of the 20 to 40 percent required to obtain a mortgage.

While there are advantages in this type of land purchase —letting you get your farm and get started farming—most states offer very little protection for the buyer who has signed a land contract. If you fail to meet your installments on time, the seller may be able to evict you. Usually there is no arrangement in the contract for prepayment, so in times of limited farm income your risk is heightened. In some states you will not only be evicted for late payment but will also forfeit all the capital improvements you have made, including the crops you have planted. Unless the seller is a relative, or someone you know and trust, be extremely careful about entering into a land contract. Get a competent attorney to write in safeguards which will reduce your risks. Alabama Farm Bulletin #400 ("Getting Established in

Farming with Special Reference to Credit") suggests the following safeguards:

1. Small payments over a long period, perhaps as long as thirty to thirty-five years.
2. Prepayment privileges that allow the buyer in high income years to make payments in advance of his repayment schedule.
3. A provision allowing the buyer to convert the contract to a mortgage after a certain amount has been paid.

Once again, be careful about land contract deals even though they are tempting offers to get you started in the country more quickly. Be sure the contract offers *real progress* toward your goal of owning and operating your own family farm.

Taking over the Seller's Mortgage

In certain instances it may be to your advantage to buy a farm or country property "subject to the existing mortgage," or by "assuming the existing mortgage." In times like these, when money is scarce and inflation is rampant, you may come across a farm where the seller is still paying on a thirty- or forty-year mortgage at 5½ to 6½ percent interest. Obviously, if you could take over that mortgage without an increase in interest, you would realize considerable savings. You would also realize initial savings in taxes and bank charges.

There are some hitches to buying property "subject to the existent mortgage." You will have no legal claim to the property until the seller has paid off his note. Theoretically, you could make your payments to him faithfully and on time, but he might not pay the lending institution that holds his mortgage. If he should default on his loan, you would lose everything you put into the property. The same sort of disadvantage falls on the seller. Theoretically, he could sell you the property and move away. You might default on your sales agreement with him, causing him to lose the equity he has built up in the property. Another thing to be wary of is that the seller might try to misrepresent the amount of interest he is paying, raising it and pocketing the difference.

By "assuming the existent mortgage" you can overcome most of the disadvantages mentioned above. However, the lending institution that holds the mortgage may have a clause in the original agreement which allows him to raise the interest to the going rate.

In order to avoid all the pitfalls in taking over an existing mortgage and to reap all the benefits, you will be wise to hire a competent attorney who has done this for many previous clients and knows all the ins and outs of such a transaction.

The risks are probably most diminished when the seller is a close relative of the buyer, has faith in his ability to make the payments, and wants to give him a "leg up."

Buy or Rent?

Now that you've worked out a long-range credit shopping list, or capital needs budget, and checked out the various credit sources for mortgage money, you may discover that your plans for buying a farm have to be altered. While buying a farm does give you equity and the right to make improvements and other management decisions, these rights of tenure can be very costly for the beginning farmer. If you can't raise the money, you will have to consider some other alternative.

One possibility, of course, is for you to keep your present city job or find another one closer to the farm. This may allow you enough income to purchase the property, farm it part time, and work in the city until you have purchased enough capital resources and built up your equity enough to allow you to farm full time. Or maybe you will find it best to lease a farm or piece of farm property and farm it part time while you keep a city job. This plan may let you gain needed farming experience and build up a decent inventory of farm equipment and livestock. If you lease with an option to buy, this program may provide an excellent path to your goal of owning and operating a farm while at the same time keeping your family income within an acceptable comfort zone.

Another alternative you might wish to investigate is one which is practiced by many beginning farmers. If you can't afford the initial costs of buying a farm that's large enough to support your family's income needs, you may want to purchase a small- to medium-sized farm and rent enough additional acreage to make your farm operation economically efficient. The added income from your rented land may be enough to allow you to make capital improvements on the land you own.

Problems and Disadvantages of Renting

Whether you rent an apartment in the city, a farmstead in the country, or just a few acres of country property, your biggest thorn is likely to be the landlord. Landlords seem to have a way of doing exactly what you *don't* want them to do. Or they demand you do things that are contrary to your wants and needs. Finding a good landlord who has faith in your farming plan and who is willing to contribute a fair amount of his resources to make the plan work will be your first rental problem. Many farmer-tenants will tell you that such a man doesn't exist. As you scour the countryside searching for him, don't forget to check with your county agent, neighbor farmers, and the local farm co-ops, and farmer groups like the Grange or Farm Bureau. If you are diligent and resourceful, you should have the same sort of success as the Mounties, who always get their man! Once you find him, have some serious discussions on what he wants for his farm and what he wants from his tenant. Find out if he has a long-range plan for his farm and if he will go along with your operating plans and management decisions.

As important as the landlord is the farm itself—its size and productivity. You and the landlord will be looking for the highest possible farm income per year. You won't be able to realize this if a large percentage of your costs and labor have to be invested toward making improvements in the land and repairs to the buildings and equipment. Nor will you be able to generate farm income if the size of the farm makes it inefficient for the type of farming you have planned to do.

Having a good farm rental situation is like having a good three-legged milking stool. It's of little use to have a good landlord and a good farm unless you get that all important "third leg," a good lease. In order to work out a fair and reasonable lease, suitable to both parties, I suggest you do a little homework. Find out all you can about the most common types of leases. These are: the cash lease, crop-share lease, the livestock-share lease, the labor-share lease, and the crop-share-cash lease. The following publications should be helpful: "Your Farm Renting Problem" (USDA Farmers' Bulletin #2161); "Your Farm Rent Determination Problem" (USDA FB-2162); "Your Farm Lease Checklist" (USDA FB-2163); "Your Farm Lease Contract" (USDA FB-2164); "Your Cash Farm Lease" (USDA Miscellaneous Publication #836); "Your Livestock-Share Farm Lease" (USDA Misc. Pub. #837); "Your Crop-Share-Cash Farm Lease" (USDA Misc. Pub. #838); "What You Should Know about Farm Leases" (California Agricultural Cir. #491).

By all means, *spell out the terms of your lease in writing!*

Nobody Ever Said It Would Be Easy

Research gathered in the northcentral region of the country shows how difficult it is for beginning families to get a start in farming without substantial family help.

"Getting established as a farm operator means more than getting started. It means achieving security of tenure on a farm with an adequate volume of business, exercising a major degree of managerial control, and owning a controlling equity in the farm-operating capital. As farms grow larger, more mechanized in operation and more specialized in productive organization, the problem of getting started and established is largely one of meeting higher requirements in land, capital, and management. One-man farms may easily require a tenant investment, though not as net worth, of $15,000 or more.

"Finding an adequate farm is a primary problem for young families, particularly for those without kinship ties to the land. Most beginning farmers start either as tenants or in some operating agreement with their parents or other farm operators who are often close relatives. For this reason, the opportunities for new operators are closely associated with available family help. About 80 percent of all beginning farmers we studied received substantial family help.

"Crop-share leases are popular with beginning farmers who do not have substantial family help. They are flexible with regard to size of the initial operation, allow an independent start with a minimum of initial capital, and, compared with livestock operations, place a limited managerial requirement on the young operator.

"Some leasing arrangements make it possible for landlords to contribute part or all of the initial operating capital. Labor-share leases and father-son agreements, for example, require little or no capital from the beginning farmer, but the number of opportunities for making a start by these means is very limited. Moreover, the young man's rate of progress from such a start depends heavily on volume of business and his managerial ability. The livestock-share lease allows the landlord to make greater contributions of both capital and management than he would probably otherwise make. In livestock areas such contributions may be important.

"Any lease agreement for beginning farmers should encourage and allow an adequate volume of business, provide for an equitable division of costs and returns, and assure compensations for any unexhausted improvements that the young tenant may leave.

"Part-time farming is a possible intermediate step toward full-time farming. It can minimize the amount of land and capital required for a start in farming as an owner-operator or part owner, but progress toward full-time farming may be slow. Part-time farming can also easily become a permanent status because making the change from urban or off-farm employment to full-time farming is hard.

"Buying land is an acceptable alternative to renting when security of tenure under a lease is a problem, when capital and credit resources permit buying enough land for efficient operation, and when buying will not impair the level of operational and improvement capital necessary for efficient operation. Some young farmers can get possession of farmland with minimum down payments by using land contracts. These are low-equity transfer devices. Some desirable provisions in such contracts include adequate length of repayment period, provision for prepayments, provision for converting the contract to mortgage financing, and an adequate grace period before default procedures may be initiated.

"For beginning farm families, savings, gifts, and inheritances, borrowings (including the use of family-owned machinery and equipment), and leasing and contractural arrangements constitute the sources of farm operating capital. These families are generally inexperienced in the use of credit, depend heavily on family sources of credit or credit backing, and particularly need intermediate-term credit. Uncertainty concerning their own ability to use borrowed capital causes some young farmers to restrict their use of credit.

"The amount of capital these families need for making a start can be minimized in three ways—by shifting capital requirements to the land owner through appropriate tenure arrangements, by substituting labor for capital, and by substituting smaller annual cash payments for larger capital investments.

"Four trends likely to characterize farming in the near future are: (1) larger and fewer farms; (2) more capital associated with one man's labor; (3) further specialization in agricultural production; and (4) a growing complexity of managerial functions in agriculture." *

So there you have it, from me and from the experts. Getting your farm and getting started in farming will be a tough row to hoe, for you and your entire family, but it can be done. And it is being done each and every year, by thousands of American families, young and old, experienced and novices, full time and part time. And you can join the trek to the country. You and yours can rediscover your ties with the earth and with Mother Nature.

The ones who succeed in transplanting themselves in the good life will be those who study hard to learn all they can about the type of farming or country living they want. Then seek the advice of a fantastic array of technical experts at every country citizen's disposal to help formulate a sound, long-range farm plan, credit priority list, and capital budget.

The successful ones will follow their plans toward their individual goals with fierce persistence. They will surely make mistakes. But that happens to us all—city slickers and country bumpkins. Planning, persistence, perseverance will make you —like Thomas Jefferson and millions of Americans who came before, and after him—*proud to be a farmer!*

Other Ways to Become a "Farmer"

After all this looking at farms and studying about the complexities of getting a farm or country place, it may be that you have just about decided to give up on your idea of becoming a farmer. Maybe you have come to realize that you and your family could not give up the comfort and convenience of life in a town or city

* Franklin J. Reiss. "Getting Started and Established in Farming with and Without Family Help," Northcentral Regional Extension Publication #8, University of Illinois, College of Agriculture, Extension Service Circular #822.

for the hazards and hard work it takes to get established on your own family farm. It is probably a reflection of your good common sense that you have faced up to the hard facts involved in making such a decision. But don't feel too bad; you can find lots of other ways to become a farmer without actually working on a farm.

Agriculture is much more than just the production of food and fiber. Millions of Americans find satisfying and rewarding jobs in many of the businesses and service organizations related to farming. One fellow I know runs an advertising firm which caters to commerical farmers and co-ops who sell to the public or want to create a good public image. There are too many jobs and professions which are farm-related for me to list here, but if you are interested, you can contact your state extension service or the USDA for career information. Of course, you don't have to take a farm-related job; you can live out in the country and work in any type of city job that strikes your fancy. You can even try your hand at living in the country and writing. It worked for Ernest Hemingway and Pearl Buck.

Another possibility is to investigate the prospects for investing in agricultural land and business. I discussed some of the basics of this kind of "farming" with Albert Shaheen, an expert in the field. Al is the president of Atlantic Properties Corporation, a California firm which specializes in agricultural acquisitions for investment purposes. His company owns more than 3,000 acres of tree, vines, and cropland in the southern half of California's rich San Joaquin valley.

A native of California, Al feels justifiably proud of the tremendous agricultural production of the Golden State. He says even though a great deal of California cropland has been taken out of production due to urban expansion, crop and livestock production have both increased significantly in the past fifteen years. He says if California could be thought of as a country, instead of a state, it would rank *third* in world agricultural production!

What does all this mean to prospective agricultural investors like you? Al says the ability to produce huge surpluses coupled with the realization that world food needs have become critical in the past two to four years has made a lot of people look to agriculture as *the industry* of the next twenty years.

USDA planners believe this country may *double* its agricultural output by the end of the century. Knowing this, the current and recent administrations have negotiated trade agreements which have opened up new markets in the Communist bloc countries for our agricultural products. These are markets which have never been expanded before.

During the Nixon Administration, Congress passed laws which had the effect of dismantling the old farm price supports and acreage allotments. In the years following, farm prices have reacted more directly with the law of supply and demand in this relatively free market. Along with increased prices for farm products, there have been tremendous increases in farmland values. This encouraged many inexperienced investors to grab up all the agricultural land they could find. Many of these investors ignored the fact that in a free market, the law of supply and demand is very sensitive to the creation of surpluses caused by overproduction.

As an example, a few years ago, California almond prices increased to $1.25 a shelled pound to the grower and almond orchards were selling at an unprecedented $4,000 to $5,000 an acre. Al says it got so people were standing in line to buy almond groves! Those speculators drove land prices too high and the increased production of almonds created a surplus. In 1975 almond growers' returns tailed off to somewhere between 75¢ to 80¢ a pound. This drop in prices helped drive the selling price of almond orchards down again—to $3,000 to $4,000 an acre. Many of those overeager investors who bought land at $5,000 an acre—hoping the land values would continue to go up—were caught short of capital necessary to invest in production. Caught in an inevitable price/cost squeeze, overextended investors were forced to come up with more money or get out. As a newcomer to the world of agricultural investment, seek the services and advice of experienced investment counselors and don't expect to make a quick kill.

I asked Al if investing in agricultural land is too risky for most of us small investors? He says not necessarily, but it does mean you should take a very realistic approach when you invest. Try to learn all you can about the quality and experience of the people with whom you are dealing. Is it a reputable firm with a good financial standing? Does the company combine experience in both the acquisition and the farm management ends of the business? Check out their track record. How many successes have they had in the past? Who are the people they are making responsible for the farming and marketing of the agricultural products? Does the land syndication company intend to hire a farm management firm, or will they farm it themselves? If they are not going to do the farming themselves, find out about the history and performance of the company providing production. Do they pay their farming bills?

Last, but perhaps most important to you as an investor, find out if the syndicating company has made reliable statements and representations to you regarding write-offs. The tax laws, which became effective in 1977, have caused important changes in the write-offs permissible for those who invest in agriculture. Consult a reliable tax attorney who is familiar with the new laws before making such investments.

With careful planning and investigation, you should have no problem in becoming a "farmer" by investing in agriculture in the next few decades. However, there are risks from drought, flood, insect and chemical diseases, and poor management. You should take the trouble to deal with reputable people who are in farming for the long haul, who know there will be good years and bad years, and who will manage your farming ventures and farming capital skillfully.

Section Six

Getting Off to the Right Start

At long last you've done it! You've made your dream come true and got a farm of your own where you can be independent and work hard to build something good and lasting for your family's future. After a little initial stumbling around that goes with any new territory, you will begin to get your feet planted on the ground and have an idea of what you are doing and what needs to be done. The big thing that will become apparent right away is that there is an awful lot of work involved with managing a farm, even a part-time or subsistence operation. It will seem at the end of each day that you've never quite been able to accomplish all that you set out to do at dawn. Suddenly you will feel tired and a little desperate! It will seem like things are getting out of hand—that you'll never catch up. If that has happened, you're ready to do some serious thinking about laying out your farm work in a systematic way.

Let's Get Organized

Remember that long-range farming plan the SCS or the FHA helped you work out? It isn't going to do you any good if you leave it to gather dust in some drawer. Get it out and go over it again, carefully. It will tell you what has to be done this year, field by field, crop by crop, and with your livestock or poultry operation.

An important part of your being a good farmer is using your noodle. Make a list of all the things that should be done this year and set up a farm calendar. This should be a *working calendar*—one which you will make many minor additions to and changes on during the course of your farm year. It can begin with the first day of January, or perhaps March 1, just prior to a new crop season. Farmers with winter-wheat operations might start their farm calendar on August 1. You should be sure to start your calendar on the same date each year so it will tie in with your farm inventory records, profit and loss statements, and fiscal reports for tax purposes. Once your farm calendar is set up, you will have a better idea of what needs to be done, and when.

Your farm calendar will show you that certain times of the year, namely spring and fall, will be the peak work load periods for your farm. Many jobs, which are important but not that critical, can be scheduled for times when you are not plowing, planting, or harvesting. Livestock and poultry operations will require farm calendars suited to the specific type of operation. Dairy operators and egg producers will have farm calendars with a fairly steady work flow. Other types of livestock producers and poultry fryer or broiler producers will have calendars with peaks and valleys. If you are going to be a part-time or weekend farmer with an off-farm job, your farm calendar will make it immediately apparent that there are certain times of the year when it will be almost impossible for you to do all the work that needs to be done. In these peak periods, you will either have to take some time off from your job, delegate part of the work to other members of your family labor force, or get someone else to do it. Don't make the mistake so many failed farmers have before you and try to do everything yourself on weekends or after work. Spring plowing, preparation of the ground, or planting can't be put off. A few rainy weekends can cause you to get hopelessly behind and, in some instances, may cause you to lose the crop for the year.

It may be, when you bought your farm, that you have made a good friend or two among the neighboring farmers in the area and they have offered to give you a hand from time to time. It would be a mistake on your part to expect too much. Your neighbors will have the same peak work load periods as you. You can't expect them to be plowing your fields for you when they should be working in their own.

The Family Farm Labor Force

Unlike many city jobs, farming is a family occupation. Right from the start, you should organize your family work force. If there is just you and your wife, work together. If there are other members of the family, get them involved in the beginning. Hold family meetings. Discuss the farm calendar. Explain what needs to be done and how important it is for everyone to share in the work and help each other. Don't be afraid to delegate and spread the work load. Assign as many tasks to family members as common sense will allow. Don't let youthful enthusiasm cloud your thinking. Assign regular chores and specific jobs to youngsters who have the capability of carrying them out. Don't create work schedules that cannot fit into an equitable overall schedule of schoolwork and social activities. Don't assign tasks where inexperience or lack of physical strength will almost assure failure. *Create winning work situations for each family member.*

Don't assume someone knows how to do a job—be absolutely certain. Nothing is more disheartening to a youngster than for dad to be constantly redoing things he tried to do, and failed. If that happens, very often, you are the one at fault. In the next family meeting, discuss your problems and reassign tasks so they can be carried out more successfully. Having regular assignments and regular meetings for family members to report on the progress of their work and the complications they are having will create a sense of sharing and pride. These can be informal meetings around the kitchen table at suppertime. The sense of responsibility that your youngsters will gain can be helpful to them for the rest of their lives.

A Daily Work Schedule

Certain tasks, especially chores connected with livestock management, have to be done according to a regular schedule or you will affect production adversely. Animals are funny that way; the cows have to be milked and the chickens have to be fed on a time schedule or you may end up with half the milk and eggs you expected. My friend Joe Daley says the Irish, who are known for their drinking prowess, have used a certain amount of "superior intelligence" to solve this problem. Irish dairy farmers have set up their milking schedules at 10:00 A.M. and 10:00 P.M. Joe says no self-respecting Irishman with a hangover is going to get out of bed at 5:30 in the morning and milk cows!

In your family meetings, set up a daily work schedule. This will help insure that tasks assigned to the various members of your family are done on time.

Plowing Should Be Done Right

Men have been plowing America's farmland for more than three hundred and fifty years now and the results have not been all that positive. The American Indians used to say that the white man was crazy. That any fool should know better than to cut into the heart of Mother Earth and kill her.

In a way, the Indians were absolutely right. Every time you plow your fields you will be disturbing the living organisms and the natural fertility in the soil. Plows tear up the nutrients that Mother Nature has taken centuries to build up in the soil and lay them bare to the effects of erosion by wind and water. Each year, millions of cubic tons of precious topsoil are lost because of improper plowing practices.

In the past two or three decades, farmers have learned more and more about conserving the fertility of their fields and about the proper ways to plow them. Sometimes inexperienced but well-meaning beginning farmers can reverse the process of years of good topsoil management by plowing fields in the wrong direction or at the wrong depth. Before you do something dumb like this, talk to your county agent or to experienced farmers and find out how it should be done. Don't make costly early mistakes. If you don't know what you are doing, ask. Or have important farm tasks custom-hired-out to competent operators so you can learn.

Custom Hiring

For the part-time farmer or the beginning operator who doesn't have a great deal of capital or experience, it makes good sense to have the big, infrequent tasks like plowing and harvesting done by operators who specialize in custom hiring work. Ask around and get the advice of successful farmers in the neighborhood about the right man to employ for this kind of work. Contract such work well in advance of when it needs to be done, or you may find that the operator you want is all booked up. Find out if he is paid by the field or by the hour. Get bids, or at least an estimate of how much it is going to cost you to have a job done. Keep a record of when such tasks are done each year and how much it cost you.

Tenants Anyone?

Even if you just bought your farm or country place to use as a vacation home, you might want to investigate whether it would be worthwhile to have a tenant farmer or manager live on the property and farm it. If you decide to do this, be careful to hire someone with experience and references which indicate he will use proper management techniques and not just ruin your farm equipment and exploit the fertility of your land. Check him out thoroughly and discuss his farming plans in full before you hire him or sign any contracts. Become familiar with the farmers' bulletins on farm rent determination listed on page 241.

If you decide you don't want to employ a tenant manager or rent your farm, it may be that you will want to lease part of your land to a full-time neighbor operator who will use it for crops or pasture. If so, you will want to become familiar with the publications on the various types of leases listed on page 241.

Keeping Necessary Records

By now you have learned that the pencil if not mightier than the plow is at least as important. Keeping good business records will insure that you have necessary information to make the best possible decisions for improving your operation and increasing your profits. Good records will help you plan ahead, check your operation on a weekly and yearly basis, and make any needed adjustments when new conditions or situations come into play. Purchase records, livestock and equipment inventories, seed, fertilizer, and crop storage information, balance sheets, profit and loss statements, cost-return analysis data and income tax information are all important tools in a successful farm business. Contact the Superintendent of Documents, U.S. Government Printing Office, Washington, D.C. 20402, for USDA Farmers' Bulletin #2167, "Family Farm Records," or write your state cooperative extension service to obtain sample forms and specific information on the types of records best suited for your farm operation.

"Special Information for Self-Employed Farmers" is a pamphlet put out by the U.S. Department of Health, Education and Welfare which explains the procedures for reporting your income for social security. It can be obtained for 5¢ from the Superintendent of Documents. Ask for OASI-25d.

Information on the types of records necessary for reporting state and federal income tax can be obtained from the USDA or by writing to the regional office of the Internal Revenue Service.

To Market, to Market

The many ways to market farm products are too specific and sometimes too sophisticated to go into in detail. As a beginning farmer you will pretty much want to rely on the marketing advice of successful local producers of your type of crops, livestock, or poultry. Don't wait until a month or so before harvesting to investigate the various options open to you. Consult your county agent, state extension service, and various local marketing cooperatives.

If yours is a subsistence-type of operation or a country place with a fairly large-sized family garden, marketing your farm products won't be much of a problem. Most of the crops and produce you raise will be used for family consumption or to feed your own livestock and poultry. However, if you are managing a full-time commercial-sized farm, the marketing phase of your operation will be one of the key factors to making your farm-business successful.

If you raise crops, you will probably sell your grain to a local elevator. The elevator may be privately owned and operated, part of a chain, or owned by the local farmers' cooperative. You will probably try all the places available to see who will give you the best price.

The price of grain and other farm products is primarily influenced by futures trading on the big commodities markets like the Chicago Board of Trade. Most individual farmers, unless they are extremely big operators, will have little reason to become directly involved in the commodities market. However, like most farmers, you will want to become familiar with current market quotations and keep a close eye on them as a kind of barometer on whether you should sell your grain right after harvest or stack it in the fields, place it in storage silos or commercial storage facilities until the price goes higher.

The average livestock producer will also use the commodities quotations as a tool to help him decide what is the best way for him to market his products. Let's use as an example a rancher with a cow-calf operation. He basically has two ways to go. He may sell his calves as soon as they're weaned to an individual operator who has a feeding ranch. The feeding ranch operator is a middleman or intermediate step in the marketing process. He will feed the young calves for a year and then sell them as yearlings. The second option open to the cow-calf rancher is to sell his calves to a feed lot, where they will be put on growing rations. What he does depends on the size of his calves when he sells them, how long he holds them, etc. If, for example, he bought his steers at 500 pounds and fed them on grass, he might sell them at anywhere from 650 to 900 pounds—or he might take them into the feed lot and feed them himself. It all depends on the pulse of the market that he gets from hearing the commodities quotations. He uses that pulse as well as cattlemen's association newsletters and various other publications to help him in making his marketing decisions.

Fruit and vegetable truck farmers often market their products through farmers' cooperatives as do poultry producers. Dairymen's associations and milk producers' groups provide the same service for their members. You will have to investigate to determine your best marketing options. Often, small-scale fruit and vegetable growers will sell their produce directly to local markets or to farmers' markets. Sometimes they sell their crops directly to canners or other food-processing firms. California and New Jersey growers often contract their entire crops to the big canning firms. California flower growers often air-freight their products directly to the eastern flower markets, a very fast and efficient way to market perishables.

If your farm happens to be on a well-traveled road, not too far from a large town or city, you may find a ready market for vegetables, eggs, small fruit, etc. by operating a roadside stand. I know one farm wife who put five of her children through college on the tidy profits from just such an enterprise.

Put Visitors to Work

After you get your farm, you'll be surprised to find that many of your friends, relatives, and even distant acquaintances from the city, who have a secret case of "farm fever," will be trekking out to your place on weekends, vacations, and every other chance they can get. If you don't want these well-meaning interlopers to disrupt your farm schedule, put them to work! My Uncle Bob and Uncle Art always did this with visitors to their farms and it must have worked because they always seemed to come back for more!

Take a Vacation

Farming has a way of being an all-consuming way of life and a seven-days-a-week occupation. You should remember that it can easily become drudgery for family members if you don't lay out your work well in advance and plan for vacations and holidays from time to time. Perhaps you will be able to make arrangements with a neighboring farmer to watch your place and do the necessary chores. You can do the same for him when he goes away. Be sure to leave a phone number where you can be reached at all times in case of emergencies on the farm.

Keep a Farm Diary

In many ways, running a farm is like being captain of your own ship. You will be wise to keep a logbook or farm diary with entries and notations summarizing the day-to-day operation. Be sure to record any exceptional occurrences, like the birth of a calf, hiring a custom operator, accidents, and emergencies. Don't forget the funny things that happen or the happy occasions. These will make good reading later, especially when you get discouraged. Thomas Jefferson kept such a diary and it is very informative for farmers as well as for historians. In future years, your farm diary will be a good way for your children, and their children, to learn the story of their family.

Take Stock Before You Begin a New Year

No matter how carefully you plan and how hard you try, it's only human that you will make some mistakes running your farm. In order to avoid making the same mistakes over and over again, it's important that you stop and take stock before beginning a new year. Use your farm business records, your family discussion sessions, and what you have learned from experience to evaluate your first year of farm management. Find out which of your farming practices were profitable and which ones should be eliminated. Don't hesitate to listen and take the advice of other family members, especially those who are studying modern agriculture in college or high school. They may have good ideas for substituting more profitable crops or new labor-saving and money-saving management techniques for ones you have been using. This is the time to adjust your plans so your operation will be able to grow instead of just standing still.

Reading and Schooling

I've been told that a doctor has to spend one day a week reading in order to learn about the latest techniques and advances in medicine which he will want to use in his practice. Advances and improvements in farm technology are equally as important to farmers. Making a living off the land is an educational experience. There are books and publications published for operators in every phase of farming. *The Farm Journal, The Prairie Farmer, California Farmer, Beef, The Stockman's Journal* are just a few of the typical publications which will be invaluable to the farm operator and which will be invaluable for keeping him abreast of the latest developments in his type of farming.

Your state cooperative extension service will have many bulletins, pamphlets, and circulars which have been prepared to help you improve your management practices, increase your profits, and fight the many diseases which threaten crops and livestock. There are also plenty of correspondence courses, seminars, and adult instructional programs open to farmers. Take advantage.

If you have youngsters who plan to go into farming, get them off to the right start by encouraging them to join 4-H, FFA, and to enroll in high school or college courses in agriculture. Books like this barely scratch the surface of information on what is happening in agriculture. Keep up your reading and schooling. By becoming a well-informed farmer, you will become a better farmer.

Expanding the Farmstead

As soon as you begin to make a profit, it will be tempting to look for ways to expand your operation. Maybe a bigger barn and more livestock. Or perhaps a complete remodeling of the farmhouse and installing a swimming pool? Sounds great, doesn't it? *Be careful.* You don't want to fall into the same sort of rat race you just escaped from when you came to the country. I'm talking about "the-living-beyond-your-means-and-deficit-financing" rat race.

Before you put your expansion plans into operation, go over them carefully. See how they fit together with the long-range farming and long-range credit plans you developed with the technical assistance people and your banker. What about your capital reserves? You may have had a good year this year, but what if things don't go quite as well next year? Will you be able to get through the next growing season if an emergency comes up? If you, your family, and your team of technical advisors all feel positive about your ability to expand and pay your way, it's probably a good idea.

Another aspect of farm expansion that needs consideration is whether the results of the improvements will produce farm and family income. If you are to continue to have a sound operation, income-producing projects should be given first priority.

Finally, you should get together with your family and try to determine whether expanding your operation into a modern commercial farm is one of your primary goals. Will it improve the quality of your lives? Or is this the time in your life when you want to use any extra money to pay off the debts you have already accumulated and invest your time and energy in your children's future, their education, marriages, and in your enjoyment and contentment? Whatever you decide to do, do it with the same thoughtful consideration you have given to all your decisions so far.

Your Farm Family—Pulling Together

The life you and your family build together in rural America can be full, enjoyable, and rewarding. If you feel good about what you have already achieved and can't wait to get up in the morning to see what the new day will bring—then you've already found a big part of what you came farming for.

But getting back in touch with the earth and the "growing world" is only part of the whole process of finding family fulfillment. You and your wife should look for ways to strengthen the family bonds and build the individual talents and character of your children. Get the kids into 4-H and Future Farmers of America. Encourage them to participate in music, plays, sports, and other worthwhile youth activities. Your wife and the girls in your family could look in on the old folks who live down the road to help with their meals and cleaning. You and your boys could help a nearby farmer when there's been an accident or during times when he has a heavy work load. There's a lot of satisfaction that comes from being a good neighbor. The entire family should take part in church, farm, and community activities.

Get to know everyone by his first name and his good works. Try to have everyone know your family in the same way. By becoming good citizens, you will all add to the quality of your lives.

Add to your family's enjoyment and sense of satisfaction by relearning some of the age-old arts associated with country living—canning, quilting, hunting, fishing, watermelon-eating, and whittling are just a few of these wonderful pastimes.

Try to find rewarding farm projects for each member of the family. Through 4-H and FFA, many boys and girls raise prize animals or sew and can for prizes and money. Many farm wives raise chickens for "egg money" or care for a kitchen garden for home consumption and for extra money. Lots of family vacations and college educations are financed by farm gardens like these.

By planning and working together, you and your family have all come a long way toward reaching your goal of the "good life" in the country. And there are years and years of fresh air, rich, clean earth, singing birds, and happy harvests ahead. By pulling together I know you will all enjoy every one of them with the good Lord's helping hand.

"The final crop of any land
is people . . . and the spirit
of the people."
—Robert Flaherty, *The Land*

Addresses of State Extension Service Directors

ALABAMA: Auburn University, Auburn 36830

ALASKA: University of Alaska, College 99735

ARIZONA: University of Arizona, Tucson 84721

ARKANSAS: P. O. Box 391, Little Rock 72203

CALIFORNIA: 2200 University Avenue, Berkeley 94720

COLORADO: Colorado State University, Fort Collins 80521

CONNECTICUT: University of Connecticut, Storrs 06268

DELAWARE: University of Delaware, Newark 19711

DISTRICT OF COLUMBIA: Federal City College, Washington, D.C. 20005

FLORIDA: University of Florida, Gainesville 32603

GEORGIA: University of Georgia, Athens 30601

HAWAII: University of Hawaii, Honolulu 96822

IDAHO: University of Idaho, Moscow 83843

ILLINOIS: University of Illinois, Urbana 61803

INDIANA: Purdue University, Lafayette 47907

IOWA: Iowa State University, Ames 50010

KANSAS: Kansas State University, Manhattan 66504

KENTUCKY: University of Kentucky, Lexington 40506

LOUISIANA: Louisiana State University, Baton Rouge 70803

MAINE: University of Maine, Orono 04473

MARYLAND: University of Maryland, College Park 20742

MASSACHUSETTS: University of Massachusetts, Amherst 01003

MICHIGAN: Michigan State University, East Lansing 48823

MINNESOTA: University of Minnesota, St. Paul 55101

MISSISSIPPI: Mississippi State University, State College 39762

MISSOURI: University of Missouri, Columbia 65202

MONTANA: Montana State University, Bozeman 59715

NEBRASKA: University of Nebraska, Lincoln 68503

NEVADA: University of Nevada, Reno 89507

NEW HAMPSHIRE: University of New Hampshire, Durham 03824

NEW JERSEY: Rutgers—The State University, New Brunswick 08903

NEW MEXICO: New Mexico State University, University Park 88070

NEW YORK: New York State College of Agriculture, Ithaca 14850

NORTH CAROLINA: North Carolina State University, Raleigh 27607

NORTH DAKOTA: North Dakota State University, Fargo 58103

OHIO: Ohio State University, 2120 Fyffe Road, Columbus 43210

OKLAHOMA: Oklahoma State University, Stillwater 74075

OREGON: Oregon State University, Corvallis 97331

PENNSYLVANIA: The Pennsylvania State University, University Park 16802

PUERTO RICO: University of Puerto Rico, Rio Piedras 00927

RHODE ISLAND: University of Rhode Island, Kingston 02881

SOUTH CAROLINA: Clemson University, Clemson 29631

SOUTH DAKOTA: South Dakota State University, Brookings 57007

TENNESSEE: University of Tennessee, P.O. Box 1071, Knoxville 37901

TEXAS: Texas A & M University, College Station 77841

UTAH: Utah State University, Logan 84321

VERMONT: University of Vermont, Burlington 05401

VIRGINIA: Virginia Polytechnic Institute, Blacksburg 24061

WASHINGTON: Washington State University, Pullman 99163

WEST VIRGINIA: Mineral Industries Building, West Virginia University, Morgantown 26506

WISCONSIN: University of Wisconsin, Madison 53706

WYOMING: University of Wyoming, Box 3354, University Station, Laramie 82071

Index